The House Of The Sphinx

SARAH WISSEMAN

HILLIARD HARRIS

HILLIARD HARRIS

P.O. Box 275
Boonsboro, Maryland 21713-0275

First Edition-2009
ISBN 1-59133-295-8
978-1-59133-295-4
Book Design: S. A. Reilly
Cover Illustration © S. A. Reilly
Manufactured/Printed in the United States of America
2009

For Chip and Ann—good friends gone too soon

Chapter One

Sunday morning, September 5, Cairo

"Soldiers, forty centuries of history look down upon you from these Pyramids."
(Napoleon, 1798)

The camel lurched to its feet, spitting and snarling.

Lisa Donahue gasped as the animal rose on its rear feet first, tipping her forward towards the hard asphalt. She gripped the horn of her saddle with both hands. Hugging the camel hard with her knees, she averted her eyes from the piles of stinking dung and straw just below her feet and concentrated on the mangy tan hide between her legs.

Her ten-year-old camel driver grinned at her, showing three missing teeth. "Is good?" he asked in his almost non-existent English.

"So far," muttered Lisa as she sat up straighter. To her left loomed the Great Pyramid of Khufu, built as the funerary castle of a king about 2500 B.C. from some two million enormous limestone blocks. Behind it were the smaller pyramids of Khafre and Menkaure, all three tombs remnants of the original Seven Wonders of the World.

To her right was modern Cairo, whose buildings had crept right up to the Giza plateau like a metastasizing tumor. It was a messy skyline, with unfinished buildings adorned with naked, metal reinforcement bars and incongruous satellite dishes. Only an hour ago, their travel guide had told them that no one finished construction in Cairo, because then they would have to pay taxes. Better to build two floors, move in, and leave the third floor as a bird roost and a porch for drying clothes and storing sesame rushes.

It was the second day of a ten-day tour of Egypt arranged through Globus and its Egyptian counterpart, MISR Travel. Two days in Cairo, a flight south to Aswan and Abu Simbel, four days on the Nile, and then back to Cairo and home. Lisa hoped she'd like at least a few of the forty other Americans traveling with her.

Beside her, James Barber perched on his own camel, clutching his ratty old hat with one hand. "The one thing we have to do in Egypt, right?" His green eyes gleamed with an unholy glee Lisa rarely saw in him. Her physician husband hardly ever got away from his Boston hospital and he hadn't had a real vacation for two years.

"So, does this trip count as a honeymoon?" asked Lisa as her camel undulated across the asphalt.

"Sure it does. Beats sitting around in your father's house freezing to death," teased James.

When they'd finally decided to marry, their incompatible work schedules had permitted only a weekend at Raymond Donahue's house in Chatham the previous November. Although they'd had the privacy Lisa craved, her Dad had forgotten to tell her the oil delivery would be late. They had pulled out every blanket in the house, built a large fire, and boiled their fresh lobsters over a roaring blaze in the living room.

"You have to admit the lobsters and lemon butter were scrumptious," she said.

"Yeah, and dessert was fine, too," said James. His reminiscent leer was spoiled by the way he waggled his bushy eyebrows at her.

Lisa felt herself blush and he laughed at her.

She faced away from her husband. He could always do that to her. Her face changed color with every emotion, and James was an expert at riling her, embarrassing her and reminding her of private moments in public settings. It did Lisa no good to complain; James said half the fun was to get a reaction, and baiting her was so delightfully easy.

The small, barefoot boys led the caravan of tourists along the road behind the pyramids. Lisa took a deep breath of Cairo's prevailing aromas of diesel and dung and relaxed her shoulders. How had the tourist police ridden their camels around the Giza plateau? She remembered seeing one guy galloping by, his legs stuck straight out in front of him, comically bouncing on the camel's hump but miraculously staying on. Obviously, one didn't grip with the knees on this mount. She tried loosening her hips and just flowing with the ungainly stride of the camel and suddenly it was easier. Rather like a yoga stretch—relax when it hurts, breathe through the stretch, and it will get better.

The ditzy ladies from Iowa ambled along on their camels just ahead. Four sixty-something women, hooting and giggling. Nancy, Bertha, Patsy, and Linda traveled on an exotic girls' outing without their husbands every year. Two of the four wore their hair teased into impossible beehives, with long, spider-like false eyelashes and thick makeup. Their skins resembled sun-baked riverbeds. Nancy and Patsy sported low-cut lacy blouses, numerous rings on their fingers, and gold chains around their over-tanned, scrawny necks.

Lisa and James had booked the Egyptian tour as an escape from their two jobs—hers at Boston University's Museum of Archaeology and History, his at Beth Israel Deaconess Hospital—and their two kids from previous marriages. Of course, Lisa had brought some work along—a series of articles she was writing: one for her museum's magazine to accompany the upcoming exhibit on "Egypt: Past and Present," and the other a more scholarly treatise on sphinxes in Near Eastern art.

Travel in a tour bus, Globus style, was about what Lisa had expected. The tourists were herded like sheep around the major sites, begged for baksheesh and American dollars at every turn, and fleeced by the local merchants. Lisa wasn't sure how many of the stall owners forked over a cut of the proceeds to their guide, Mahmoud Sadaat, but Mahmoud certainly seemed to know all of them.

"Ramses," said her young guide, pointing at the camel.

Lisa laughed. Only in Egypt could a mangy old camel be named after a pharaoh. She pointed at the boy and raised her eyebrows.

"Ali Baba," he said proudly.

"And the forty thieves," added Lisa to herself, thinking of his extended family of brothers, uncles, and cousins who probably all made their livings off tourists like herself.

"How does it feel?" asked James, moving up and down on his mount next to her.

"Like I'm unhinged at the hips," she said. "But okay."

An older man wearing a dirty gray gelabiyah, the traditional male Arab's full-length robe, motioned to the boys to halt their camels so he could take pictures of the tourists. James struck a silly pose with both arms spread wide. Lisa supposed she'd pay the price for the finished photo, if only to prove to their friends back home that they'd really ridden camels. Of course, that was what Ali Baba's relatives were counting upon—they'd be able to feed their families today from the fees collected from this one busload of American tourists.

Fifteen minutes later they were back in the grubby alleyway where the camels parked when not giving rides to tourists.

"Lean back, Madame," said Ali Baba urgently.

Obediently she leaned backwards and her stomach muscles tightened as the camel lurched downwards, folding its front feet under it while bringing Lisa's knees alarmingly close to the hard asphalt.

A scream tore through the air to her right.

Dismounting quickly, Lisa looked around.

Bertha Miller, the plumpest of the ladies from Iowa, lay crumpled on the pavement next to her camel. "My arm!" she shrieked. "My arm!"

Lisa ran over, joining a gathering crowd of horrified tourists and anxious camel drivers. James passed her, reaching the injured woman at the same time as a heavy man in his seventies.

"I think it's broken," whimpered Bertha. Her face was as white as skim milk. "I fell off because I was so dizzy all of a sudden."

James knelt on the hard road surface, narrowly missing a pile of camel shit. "Can you move your arm at all?" he asked, turning her wrist back and forth gently.

"It hurts a lot when I try to bend it," Bertha said, tears running over her plump cheeks. "And I'm so hot."

The older man, who identified himself as retired New York physician Stanley Rosenberg, felt Bertha's forehead and frowned.

"I think she has a temperature as well as a broken arm," he said.

Stanley helped James clear a space for her to stretch out. Someone produced a folded towel as a cushion for Bertha's arm and another for her head. Lisa marveled at James' quick reactions until she remembered his experience as a young doctor with Doctors Without Borders in Tanzania.

A young woman from the back of the bus produced a painkiller and some bottled water, and James gave both to Bertha. Then he stood up to talk to their guide, Mahmoud Sadaat. Mahmoud was a portly Egyptian with thinning dark hair and shiny brown eyes.

Bertha's friend Patsy Davis bustled over. "Hasn't Mahmoud phoned the ambulance yet? Mrs. Miller is in agony!"

"Take it easy, Patsy. I saw him—he phoned right away," said Lisa, resenting Patsy's bossiness. But, she added silently, who knows how long it takes to respond in Cairo traffic. She remembered that her guidebook described driving in Cairo as like the chariot race in the movie version of Ben Hur, but without the chariots. It was a very apt description. Cairo drivers showed no concept of lanes and drivers clearly preferred horns to turn signals.

"Bertha's doing okay," said a tall, red-haired woman Lisa had noticed on the bus earlier. She was kneeling on the ground holding the hand of Bertha's uninjured arm. Lisa saw her take a pulse and asked, "Are you a nurse?"

"Yes," said the woman with a crooked smile. "Betsy Grover, the world's most traveled ER nurse. Actually, I'm from Detroit. You're husband's a doc, right?"

"Yeah, a radiologist. He's right behind you," Lisa nodded at James, who was still talking to their tour guide. "He's James Barber, I'm Lisa Donahue."

"Pleased to meetcha," said Betsy. She turned back to Bertha. "Hang on, honey, the ambulance is on its way. Has the Ibruprofen started working yet?"

"A little," mumbled Bertha. "It's not so bad as long as I don't move my arm."

James and Stanley Rosenberg encouraged everyone except Betsy and Patsy to step back so Bertha could get some air. The tourists milled around, chatting and sipping from their water bottles, waiting for the ambulance.

Twenty minutes later, Mahmoud appeared at Lisa's elbow. "I can't understand," he said. "I telephone quickly." He looked anxiously at the stricken woman, and then at his watch. Lisa sympathized with him—what a thing to happen at the beginning of a tour! Mahmoud Saadat was not the most efficient tour leader Lisa had ever met, but he was pleasant enough. His English was passable, and his history and Egyptology training quite good.

Another officious tourist, a balding forty-something man from Chicago (or was it Boston?), ambled over. "Why isn't the ambulance here yet?" Lisa couldn't see his nametag, but she thought his name was Wallis.

"Cairo traffic," said James. He leaned over and whispered to Lisa, "And maybe they don't have as many ambulances as we do at home."

"Oh, we have plenty of medical vans," said Mahmoud, who'd overheard them. "But the staffing, that is another matter."

"What about all those guys lurking at the end of the alley? Shouldn't we get them to move?" said Lisa to James. She felt a need to shield Bertha from the curious eyes of the camel drivers and their relatives who seeped out of the nearby mudbrick walls.

James glanced around. "They're doing no harm," he said. "And we might need some local help if the ambulance doesn't arrive soon."

As he finished speaking, a small white van appeared at the end of the alley.

"Finally!" said James. "Hey, Lisa, see if you can make the other tourists move out of the way."

"Sure," said Lisa. She urged people to move back—again—and most of them complied. Together, she and James cleared a path for the emergency medics.

Off to the right, Lisa saw their "suit"—the security guard who traveled with the tour group everywhere—in a huddle with a white-uniformed tourist policeman and an adult camel driver. The guard—Zawi Something—had initially blocked stray cars from entering the alley where Bertha lay; now he and the policeman and the camel driver faced away from Lisa. She couldn't see Zawi's face, but she could see the face of the camel driver. He held a cigarette in one hand and a small package in the other. She was just close enough to see that it was wrapped in brown paper, tied with string and marked with Arabic and a small red symbol. Cigarettes? Or maybe drugs? Lisa chastised herself for thinking in Hollywood terms. She couldn't see whether it was Zawi or the policeman who accepted the little package.

The commotion of the stretcher-bearers getting into position around Bertha Miller and sudden awareness of her almost bursting bladder cut off her train of thought.

Ignoring her own discomfort, Lisa watched as Bertha was lifted by four medics. The poor woman, surrounded by her ditzy friends, yelped with pain as her arm was jostled on the way to the ambulance. Mahmoud hustled all the tourists back to the bus and then returned to give instructions to the ambulance driver. Lisa heard him mention calling the American Embassy; that meant Bertha would be in good hands.

By the time the white van pulled away and headed for the nearest hospital, it was two o'clock and the whole group was fractious and starving.

Mahmoud Sadaat took them to a Cairo restaurant surrounded by date palms and flanked by musicians at the door. One of them, a dark young man beating on a tambourine, swayed as if he were on the verge of belly dancing. His liquid brown eyes fastened onto Lisa's face and he beat his tambourine faster. Lisa smiled as she passed, enjoying the insistent beating of the drums. Part of her wanted to linger, but the rest of her body was intent on finding a bathroom. On her way to the necessary facility, Lisa passed a young, veiled woman seated cross-legged in front of an ancient brick oven. Smiling at the visitors, she lifted out circles of fresh pita bread and placed them on the surrounding brickwork to cool. Lisa remembered seeing fresh pita set out on the sidewalks—much less sanitary than these low brick walls—in another part of the city.

Lisa and James found places at a long table covered in a white tablecloth. They were joined by Betsy Grover, the ER nurse, and an Ohio family. The Reese parents were about Lisa's age, with a small boy named Nicky in tow. Nicky's engaging grin reminded Lisa of Sam.

"Hungry?" she asked him.

"Starving!" he said, and bounced on his chair. His red cowlick dropped over his small forehead.

Barbara Reese rolled her eyes at Lisa. "He's always hungry," she said with an indulgent smile.

"Yes, I know what you mean. Our son, Sam, is just the same," said Lisa.

Soon the waiters appeared with little plates of baba ganoush, hummus, and tomato salad. Before he left them to make phone calls, Mahmoud had told them to eat freely. "The food here is very good, very safe. Not to worry."

James raised his eyebrows, but Lisa needed no second testimonial. She and Nicky Reese were already scooping up garlic-laden hummus and delectable roasted eggplant with torn bits of the warm, freshly baked pita. She noticed that James, by nature more cautious around foreign food, steered clear of the tomato salad.

Patsy Davis and her friends alternately wailed over Bertha Miller's fate and caught up on their shopping triumphs.

"Phil will be devastated he's not here with her! I'll call him this evening. What's the time difference from Dubuque to here?"

"...and he only asked twenty pounds." That was Nancy, bragging about one of her purchases.

"Egyptian, or English pounds?"

"Well, it was five American."

"She's going to miss the best part of the tour!" cried Patsy.

Betsy Grover wiped her appetizer plate clean with a piece of pita. "I hope she has insurance," she said to Lisa.

Lisa hoped so, too. She was thankful she'd paid the extra sixty dollars for herself and James for emergency medical care and travel interruption. One expected things like that in Africa, or at least she did.

A coffee-colored baby at the table across the aisle squealed with delight as his pretty young mother made faces at him. Lisa's gaze fastened avidly on the bright little face. A baby. So adorable, with that big forehead and those chubby cheeks. Maybe she and James should have another child, one that belonged to their marriage alone...

Lisa was yanked back into the here-and-now by the return of their guide, Mahmoud Sadaat. His phone calls completed, he looked tired and hungry. But the ladies from Iowa mobbed him before he reached the buffet table.

"Is Bertha going to be okay?"

"Can she join us in Luxor?"

"Is she in the best hospital? I mean, for Cairo?"

Mahmoud mopped the sweat from his brow with a dubious handkerchief. "Mrs. Miller will have her arm mended. She maybe will join us later. I will telephone her tonight," he said, moving hopefully towards the appetizers.

"But Bertha can't travel all by herself! Are you going to arrange a guide to escort her?" cried Patsy.

Poor Mahmoud. He sure had his hands full with the Iowa ladies. Lisa switched her focus to the main course, a delectable mixed grill of lamb and chicken with rosemary. She might be in her mid-thirties, but she still had the appetite of a teenager.

James, eyeing her full plate, raised his eyebrows. "You might want to slow down a bit. I'm sure tourists get sick here—they probably call it King Tut's Revenge instead of Montezuma's Revenge."

Lisa made a face at him and took another mouthful of tomato salad.

She was to regret that salad—very soon.

Chapter Two

Same morning, Luxor

Seventeen-year old Abdul Farouk raced through the family compound near the Avenue of the Sphinxes at Luxor, hell-bent on beating his cousin Mohammad, Jr., to the next batch of tourists. It was a school holiday, and he was determined to make every hour pay.

His dusty olive-green gelabiyah was draped with cheap, colorful canvas bags tied together at the handles. Each was decorated with ancient Egyptian gods and goddesses copied from the tombs nearby. Abdul planned to ask for three American dollars, a clear profit for him. Each bag was manufactured by the extended family of the Farouks for a mere fifty cents apiece. Making bags was only one of several money-making endeavors, not all of which were legal.

He smiled as he ran, thinking of how he would spend the extra money. Half for his secret savings account so he could leave Luxor and study biology and medicine in Cairo, a quarter for the cigarettes and CDs he required to make life bearable, and the rest for his mama's older brother, Uncle Ali. His uncle did not realize that Abdul hid most of his earnings from the family.

Abdul slowed his steps and frowned as he thought about his mother. Mama was terribly worried about her sister Safaa, Ali Farouk's wife, who was very sick. No one else at home was ill, so Abdul couldn't see what all the fuss was about. But his mama wouldn't make her honey and pistachio pastry if she was nursing Safaa, and that impacted Abdul right where it counted—his voracious appetite.

The only person Abdul would share his favorite food with was his younger sister Zara. He slowed down even more until he was barely shuffling as he thought about Zara. She was sixteen, with a slight but seductive figure and huge brown eyes. He knew his father would choose a husband for her soon, but it wouldn't be the one she wanted.

Zara was in love with a penniless youth called Farzad Hammadi. He was just eighteen and worked in his father's alabaster stall in the souk. Farzad was Abdul's best friend and accomplice in all his ploys. Abdul's father, Zawi Farouk, was always trying to better himself and increase his income by investing in local businesses; he would choose a business partner, an older man of means, and offer his young daughter as a bribe to cement a new alliance. But Zara wanted to finish her education. She was full of fire and melodrama. What would she do when her father tried to push her into a marriage she didn't want? She...

"You're late," taunted Mohammad, who had already grabbed the best spot by the ticket booth at the entrance to Karnak.

"No matter," replied Abdul, unloading his gaudy canvas bags onto a nearby granite block. "It is early. In sha'allah, we will have rich American tourists all day, and I will make twice as much as you." He glanced disdainfully at Mohammad's shabby three-stringed instruments. They looked like they would fall apart if anyone strummed them for five minutes.

"Huh!" said Mohammad, and spat within two centimeters of the nearest canvas bag, a blue and gold number that depicted the Queen Nefertiti in gorgeous array.

Abdul yanked his merchandise out of range. "Hold your spit," he snarled. "You son of a jackal!" He added some other choice epithets in gutter Arabic.

A typical day had begun.

Zara Farouk entered the miniscule kitchen shortly after her brother Abdul had vacated it. She unwound her headscarf and draped it over the chair, sighing with relief as a slight breeze ruffled her dark hair.

Her eyes gleamed with anticipation. Now was her chance to make some pastry, while her mother and aunt were occupied, the other women were at the market, and the men were away pursuing various money-making schemes.

Zara loved to cook specialty dishes and sweets, but she rarely got the opportunity. With the women from three families crowded into one compound, sharing cooking duties, the personal desires of the youngest were rarely considered. When Zara was asked to "help," it usually meant menial tasks like scrubbing vegetables or sorting lentils or washing endless piles of dishes.

Quickly, she assembled the ingredients for a local version of baklava: phyllo pastry, butter (with any luck, not rancid yet), honey, chopped pistachio nuts, cinnamon, and cardamom. In the narrow slot between the stove and the wall, she found a flat pan and greased it. Then Zara melted the butter. She added honey and spices and cooked the fragrant mixture over low heat until it was ready to spread.

Now for the best part. She arranged the phyllo sheets so she could reach them easily, buttering each before laying it in the flat pan. Every few layers, she added nuts and a drizzle of honey. The smooth motions of buttering soothed her and allowed her mind to wander.

What was Farzad doing this time of day? Probably working in his father's shop, absorbing methods of attracting and hooking customers, storing up knowledge against the day he would run his own alabaster shop. With Zara, of course, as his wife. He would choose and acquire the best pieces and sell them while she kept the books. Maybe south, near Aswan or Abu Simbel; that would be the best way to get away from Luxor and Zara's impossible father, Zawi Farouk.

Zara paused, her buttering brush in midair as she thought of her father. Mean old Papa. He had recently made noises about marrying her off to a colleague, but she was determined to refuse any such proposal. Papa was very old-fashioned, but she didn't really believe he would force her to marry against her will. She was her father's favorite, the youngest of three daughters. But Farzad Hammadi was the love of her life, the future father of her children, her destiny. She shrugged her slender shoulders as if her father's plans could slide right off and wither on the dusty floor. Gently, she picked up the next sheet of phyllo and buttered it.

Her position would be so much better if she had a college degree and money of her own. Zara's grandmother and mother had never finished high school, let alone college. Zara was unusually good at numbers, and she wanted to be an accountant. Then, she would never be dependent on her father and could help Farzad with the business. But Papa was pushing her towards stopping with a high school degree when she wanted to go to college. She couldn't move to Cairo if she married Farzad, but there were other places with night school and correspondence courses. But her father made it so hard for her to save money. He feared she would buy cosmetics and CDs—or shoes—instead of food for the household, so he counted out only the coins she would need for each trip.

Zara looked with satisfaction at her new sandals, acquired yesterday for only one dollar American. They were pink leather, decorated with pink and white plastic jewels. The new shoes, coupled with her new pink-and-white striped headscarf, made her long skirt and long-sleeved blouse look much more fashionable.

The beautifully layered confection was ready to bake. Zara slid the pan into the warm oven and set a small timer. She was about to make herself some mint tea when her mother walked into the room.

Mina Farouk looked around the kitchen and exploded as she saw the pile of dirty dishes. "Look at this mess! I am just about to start the dinner after nursing your aunt all day, and now this!"

"I am making Abdul's favorite pastry. And I thought Aunt Safaa might like some, too."

"Safaa is in no condition to eat anything, let alone sweets! She will need my special broth and a lot of care. You should be helping me with the regular cooking, not making sweets."

"Mama, I will help you cook, I promise. And I will clean it up," said Zara, sensing her mother's distress. "How is Aunt Safaa?"

Mina sank down in the rickety chair and covered her face with her hands. "She is worse. I think the doctor must come again. I can't understand why she isn't getting better..."

"I will go fetch him," promised Zara. "You know running over to his office is often faster than trying to telephone at this time of day. And I will make the couscous when I return, so you don't need to worry about that."

And she could sneak in a quick visit to Farzad's shop on the way back...

By late morning, Abdul had made enough money to make his date-brown eyes shine. Today, he could buy the latest Mariah Carey CD in the market and still put most of his earnings into his savings account.

He slung his gaudy canvas bags over his shoulder, not trusting his cousin to watch over his stuff while he grabbed some lunch. His pace quickened and his stomach rumbled as he remembered his mother's leftover couscous flavored with almonds and apricots.

When Abdul arrived at the Avenue of the Sphinxes, there was very little traffic. At the hottest part of the day, most people retreated to the cool, mudbrick houses with their thick walls that kept out the heat. Abdul's house was part of the Farouk compound, a cluster of one-and-a half story buildings covered in mudbrick and wood. The roofless upper floor served as sleeping quarters in really hot weather as well as storage—perfectly safe from water damage as it hardly ever rained. Three of the five Farouk brothers lived in the compound, along with their families, a mangy dog named Tut, and a few chickens who pecked grain in the central courtyard.

No one was in the tiny kitchen, but Abdul smelled honey and spices and saw evidence of recent baking stacked in the sink. Zara had made baklava! Where had she hidden it? He opened the cupboard and found it—it was still warm. He grabbed a knife and cut himself a generous wedge. His mouth full, he rooted around in the miniscule fridge that was the largest his father could afford for the couscous leftover from yesterday. Good, there was a small amount left. Abdul wrapped it in a stale pita bread and bit into it before his mama could appear and ask him to save some for his sisters. It was all part of living in a tiny house with twelve people and never quite enough to eat.

Normally his appetite was huge, but today Abdul could eat only half the pita. Absently, he wrapped the second piece and stuck it into his robe. He could hear the rising and falling tones of his mother's voice in the bedroom six people usually shared down the hall. She sounded upset. Abdul lifted his chin and listened for his Aunt Safaa's reply. The soft babble reassured him; she wasn't unconscious anyhow, so it couldn't be too bad.

Glancing around the tiny kitchen, he saw that Uncle Ali's huge, azure tobacco jar containing his special tobacco had shifted slightly. It stood close to a second jar—full of ordinary, everyday tobacco—and a chair they all used as a clothing rack. He couldn't resist sniffing the aromatic mixture as he had done before; he'd only done this when his uncle, who was possessive about his stash, was absent. Abdul lifted the lid and scooped up a handful, holding it close to his nose. A faint whiff of anise—or was it basil?

There was no package this time. Two weeks ago, when Abdul had sniffed his fragrant handful, he'd discovered a small, narrow object inside. His agile fingers had quickly untied the string and loosened the brown paper to reveal a cardboard tube that was labeled in a foreign script.

Inside, he found a small glass ampule, pointed at one end. It was terribly fragile; the slight pressure of his questing fingers caused the thing to shatter in his hand, spilling fine dust on his aunt Safaa's pink headscarf lying over the chair back.

Carefully, he rewrapped the now empty cardboard tube, tied the package with string in the same pattern of knots, and thrust it deep in the tobacco jar.

Now, two weeks later, the memory made him feel slightly guilty. But Uncle Ali hadn't even noticed the missing glass container, so it couldn't have been important. Abdul rubbed his forehead, which was aching and felt hot. He was extra thirsty, too. He looked longingly towards the bed he shared with two cousins; a nap would be wonderful. But his father and uncle would beat him for being lazy. Abdul filled his water bottle and left the house for another afternoon shift of fleecing tourists at Karnak.

Chapter Three

That afternoon, Giza

"Then he also erected some huge statues and massive, man-headed sphinx figures." (Herodotus, Histories 2:175)

It towered forty feet above her, solid and sinister. The vacant eyes stared past her, the right side more smashed than the left. The missing nose only added to its visceral impact.

The Great Sphinx, built by the Pharaoh Khafre, took Lisa's breath away. Awe swelled in her throat. She'd been studying sphinxes of the ancient Near East for the past five years, but the reality was so much more than any image captured in a book or on a computer screen.

The Great Sphinx of Giza was called "Abu al-Hol," or "Father of Terror" in Arabic. Lisa thought he deserved the name. He was monstrous, huge, and timeless.

Lisa remembered that in Greek, "sphinx" meant "strangler." The Greek sphinx—a female—was a mythical creature with the body of a lion, the wings of a bird, and the head of a woman. She was an oracle, celebrated in Homer with the famous riddle asked of Odysseus: "What walks on four feet in the morning, on two at noon, and on three in the evening?" (Answer: man. He walks on all fours as a baby, upright as an adult, and with a staff in old age). She mentally composed another section for a series of articles she had promised to write for her museum's magazine:

"The Egyptian version of the sphinx retains the lion's body, but the head is that of a pharaoh in traditional headdress, or was, until someone—probably Napoleon's men in about 1800—hammered off his nose. Later, his beard fell off and part of it was carted off to the British Museum by nineteenth century travelers.

The Great Sphinx wasn't always as exposed as it is today. Eighteenth century drawings of the Great Sphinx show it buried up to its neck with sand. Over time, several attempts were made to clear the sand, but the wind always brought it back. A massive restoration project led by Zahi Hawass, the head honcho of the Egyptian Supreme

Council of Antiquities, has been conducted over the past twenty-five years. An international team studied the problems caused by pollution and the rising water table and prepared a plan for long-term preservation of the monument."

She swabbed at her face with the sleeve of her rather scratchy, sun-protective shirt and stepped back to obtain more of a panoramic view with her camera. The hollow feeling in her gut changed to an ominous rumbling. James was right. She who ate too much in a strange restaurant would live to regret it. Lisa hoped she would not be up all night, clinging to a cold porcelain throne instead of dreaming about tombs and sphinxes.

James appeared next to her. "Is the Great Sphinx everything you expected?"

"And how," Lisa said, lowering her camera and pulling out her water bottle. She unscrewed the cap and glugged down a third of the contents. One of her least favorite things about third-world travel was remembering to drink only bottled water. It was especially hard brushing her teeth, when the habit of decades dictated that she run the toothbrush under the faucet. Or in the middle of the night when she couldn't remember where she was when she stumbled into the bathroom to get a drink. She and James had decided that the symbol of America should be the public drinking fountain: clean, fresh water available everywhere—something totally unknown in most of the non-Western world.

"You'll be able to get lots of atmospheric background for your articles," James said.

"Yes, oh yes," Lisa smiled.

This was a tease, since they both knew that she'd do a good part of her research and writing in Boston University's library when she arrived back home. But research on the mythology of sphinxes in several ancient cultures of the Near East was a second excuse—if she needed one—for the trip to Egypt. Lisa had longed to visit Egypt since she was sixteen. And the popular articles on "Egypt: Past and Present" would almost write themselves when their tour group took the Nile Cruise.

That is, if she could concentrate on writing—the incident with Bertha had shaken her up. It felt like a bad omen, right at the beginning of their trip. Lisa had looked forward to traveling without Emma and Sam in tow. So why was she being a Nervous Nellie now that they were actually in Egypt? Maybe because her friend Barbara had regaled her with horror stories of her husband getting hospitalized in Nairobi after a car crash in Tanzania. Or maybe just because her stomach was so unsettled…

"Is he really male?" asked James, squinting into the sun as he gazed up at the chin of the Sphinx.

Lisa was glad to turn her attention back to archaeology and mythology. "This version is. The Greek sphinx was female, but the Egyptian one was all tied up with sun-worship—the Pharaoh as the incarnation of Ra, the Sun God."

"Yeah," said James, who was a quick study with the Lonely Planet guidebook they'd brought along. "But didn't you tell me some scholars dispute its exact origin?"

"You're right. But from what I read right before we came over here Zahi Hawass—he's the head honcho of the Egyptian Antiquities Service—and his gang have pretty much proved that the Pharaoh Khufu was Ra, at least as far as his son Khafre was concerned. The Sphinx represents Khafre as Horus giving offerings to Ra."

"But since Ra is also Khafre's father, it's sort of ancestor worship too, isn't it?" said James.

"True. But most Egyptologists would probably argue with you there. Everything—and I mean everything—in ancient Egypt had to do with the gods and worshipping them. Religion was never separate from daily living. Belief in the gods of sun and sky and underworld infused everything people did." Lisa frowned a little as a sinister cramp struck her lower gut.

"You're looking a bit green about the gills," said James, who was an expert at reading Lisa's facial expressions.

"I feel green," she confessed. "I think lunch is catching up with me."

"I'll see what I have in my bag after we get back to the hotel."

Chapter Four

Monday morning, September 6, Cairo

"Cairo is the Mother of the World." (A Thousand and One Nights)

"Listen up, everyone," said Mahmoud Sadaat after they had taken their seats and the bus was moving. "Today, we finish the program in Cairo, then we take late afternoon flight to Upper Egypt, to Aswan. We see High Dam there, very interesting, and Lake Nasser. Some of you have signed up for the Abu Simbel trip…"

Lisa gazed out the window, remembering with pleasure her glee when she'd opened her curtains that morning and seen a pyramid looming over the arched façade of the hotel. She and James had managed an early morning swim in the huge, outdoor swimming pool—also with a pyramid view. The grounds were beautifully landscaped with palm trees and flowering bushes, and they'd enjoyed a quiet breakfast in the same garden restaurant where they'd heard an Egyptian guitar player the night before playing Bob Dylan songs. The incongruity of the setting, the Egyptian accent, and the American songs had made them both smile; James had dubbed the singer "Dylan Ali."

The evening had been a far more auspicious start to their Egyptian tour than the first day with Bertha's accident and the long wait for the ambulance. Lisa's mind fclt alert and she was almost over jet lag. Her stomach, however, rumbled and roiled in a ominous fashion. She promised herself she'd be much more cautious about what she ate for the rest of the trip.

"Why are we flying to Aswan?" asked someone at the back of the bus. "I heard you can take a train down there."

"It is much safer to fly with large group," Mahmoud said shortly.

Lisa nodded, recalling that the guidebook said all large groups were handled this way since the Egyptian Tourist Police had almost no control over Middle Egypt between Luxor and Aswan. Since the massacre of German tourists at Luxor in 1997, the Ministry of

Tourism was very anxious to prevent any incident that might injure tourists. Or discourage overseas visitors from coming to Egypt and spending all those lovely greenbacks.

"Why is it called Upper Egypt, when it's south of here?" asked Patsy Davis, as they passed a canal with a dead cow in it.

"Everything goes back to the Nile River and its source in the south," replied Mahmoud, holding up one hand with fingers spread wide. "This is Upper Egypt, the highlands where is the source of the river," he said, pointing to his elbow. "The current flows south to north, until it reaches the Delta, or Lower Egypt." He pointed to his raised hand. "The Egyptians thought all rivers in other countries flow the wrong way."

"Now, I want you to know something about Cairo. It is largest capital in Arab world, over fifteen million people. The city was founded in tenth century AD by the Fatimid Dynasty."

The bus pulled up to the curb behind a long line of similar tour buses, parked in front of an imposing reddish-tan building. Mahmoud, checking his watch, said, "We are at the Egyptian Museum. We will have one hour here. At eleven o'clock, you will be back on the bus."

Lisa heard some grumbling behind her. "Like a bloody herd of cattle, we are. One hour? I hear it's a huge museum; that will never be enough time."

She agreed. She had every intention of coming back with more time, after they finished the Nile cruise and returned to Cairo. Lisa and James descended with the others, part of a milling throng of tourists from several countries. She heard guides trying to corral their groups, exhorting them in French, German, and Japanese, to line up in an orderly fashion.

The long line snaked through security. She stood for ages, watching impatiently as the men were asked to take off their belts and empty the change from their pockets. The women were luckier, since most of their belongings were in purses that could be sent quickly through the X-ray machine.

Lisa's heart beat faster as they entered the high-ceilinged, dimly lit main hall that held some of the world's most famous Egyptian artifacts. Pillars marched down the center, and between every pair was a sarcophagus, or mummy case. Here were the pink-and-black granite originals of the Pharaohs who built the pyramids—even her own Boston museum had plaster casts of them. Boston had quite a few original artifacts, but not as many as New York or Paris or London or Berlin. The European collectors had made off with an incredible amount of ancient art from the lands around the Mediterranean back in the heyday of great museums—when foreigners could transport artifacts across borders with impunity. Now, with modern legislation and a resurgence of national pride, many of the looted countries were demanding that their artifacts be returned.

"James, look at this," Lisa paused by a case showing X-rays and other evidence of the scientific study of mummies.

"Ah," said James. "One of your favorite subjects."

Lisa smiled. She really did love mummies and everything connected with them. "You can see the display philosophy is different over here," she said.

"What do you mean?"

"Well, at home, the new trend is Disneyland museums—all glitz, with minimal artifacts and lots of hands-on stuff for the kiddies."

"And very little meat for intelligent adults," said James, who shared Lisa's love for older museums just stuffed with objects. The out-of-fashion Cabinets de Curiosité—with everything from rocks and quail eggs to priceless Greek vases in the same case—appealed to both of them.

Lisa cast a professional eye over the labeling in the Egyptian cases. American labels could be very pretty, in neat fonts and lettering, but the text was often dumbed down to about a sixth grade level. Here in Cairo, the labels were informative—when you could read them. They were typed in Arabic, English, and French, and many objects had no labels at all.

Mahmoud Sadaat, who knew Lisa's background, walked with her into the King Tut exhibit. "I study here two years," he said proudly. "And my exam, it was like this: they bring me into one hall in the museum and point at a case at random, saying 'tell me about this.'"

"Good heavens," said Lisa, looking around her at the vast number of objects she knew nothing about. Even with her background in Near Eastern archaeology, she'd have had trouble with an exam like that. She noticed a few objects missing, and asked Mahmoud if they were being cleaned or restored.

"No, they go to United States on touring exhibit. But you will not need to pay money again to see that exhibit," he said with a grin. "You see much more right here, in Cairo."

As Lisa admired the gold gilt furniture and spectacular jewelry, she had to agree. There was so much here that the gaps in the collections were barely noticeable.

Lisa discovered a case full of tomb equipment: hair ornaments, cosmetic jars of alabaster, and exquisite faience ushabtis, the little statues of servants meant to serve the deceased in the afterlife. These little figurines were the best she'd ever seen. Maybe her archaeologist friend, Jeff Warden, could get her in to the storerooms when they returned to Cairo.

All too soon, she heard Mahmoud clap his hands. "Time to drive to Old Cairo," he said to the group. "We go visit the Alabaster Mosque, where Mohammed Ali is buried, then the oldest synagogue in all Egypt, and then a carpet factory."

"Carpet factory?" wailed Nancy Keller. "But I want to go to the big market, you know, the one all the tourists visit!"

Mahmoud smiled indulgently. "You mean the Khan al' Khallili. But we don't have enough time today; carpet factory is quicker, and very interesting. It is a school, and many Egyptian teenagers work there. You will have plenty of time for shopping in the big bazaar when we come back to Cairo, in five days time."

As the bus rumbled and swayed along narrow streets, narrowly missing pedestrians and vehicles of all sorts—Lisa saw as many donkey carts and bicycles as cars—Mahmoud filled them in on the history of Old Cairo.

"You have all heard of Napoleon, yes? He came to Egypt with army in 1798. In forty-five minutes, his army defeated the Mamluks—these are Turks—at the pyramids which you have seen yesterday..."

Lisa thought about Napoleon's more academic achievements: along with his soldiers, he brought a second army, composed of scholars and artists, who were assigned to record everything they saw. The result, a twenty-four volume tome in French, was still a standard reference for archaeologists and historians.

"In 1800s we have very important pasha, or governor," continued Mahmoud. This was Mohammed Ali, an Ottoman Turk. He also massacred many Mamluks. But he do good things also. He introduce cash crop of cotton and started public education system here in Egypt..."

"Education is free in Egypt, right?" interrupted David Wallis.

"Yes, that is so. Free from primary grades through university. But many graduates still do not get jobs, and many people in smaller towns cannot even read," said Mahmoud.

"Now, Cairo was transformed by Turkish rulers. They started big building program, much architecture. Some of the limestone blocks were taken from the pyramids to make the new buildings."

Lisa reflected that this was nothing new. The Greeks and Romans—in fact, most ancient civilizations—were notorious for cannibalizing stone from the temples and palaces of their predecessors. As the earliest recyclers, they probably figured why waste good stone, when it was so much cheaper to just take pieces of an existing building?

"You will see in a few minutes the city's oldest Jewish synagogue, called Ben Ezra. It was built inside early Christian church. Cairo supported all three of the greatest religions—Christianity, Judaism, and of course, Islam. We will walk there from this point."

The bus, belching exhaust, stopped on a narrow sidewalk and they all piled out. Lisa enjoyed the short walk along cobbled streets with the buildings on either side leaning inwards. The arches overhead and the window boxes full of flowers reminded her of the Old City in Jerusalem, and she experienced again the feeling of walking back into time.

James, intent on taking another photograph of an interesting doorway, dropped behind. Lisa entered the ancient synagogue and admired the dark woods and richly colored paintings. She was reminded of another church-within-a-church she'd seen in Cordoba, Spain; only that one was a mosque instead of a synagogue.

Soon they were at the entrance of the Alabaster Mosque, an imposing edifice all in white, with layers upon layers of domes, windows, and arches. In the forecourt was a fountain for the ritual washing of hands and feet and an absurdly ornate clock. Mahmoud told them it had been a gift from a king of France—in exchange for the Egyptian obelisk that now adorned the Place de La Concorde in Paris.

"Ladies, if you are wearing shorts or small tops, you must accept covering robe," said their guide.

Lisa thought she was pretty well covered up, but she was handed a green robe that enveloped her like a tent. She slipped off her shoes in company with dozens of other tourists doing the same, and stepped onto the plush red carpet. Above were numerous chandeliers, all lit, which made the gold and cream alabaster glow.

"Here we see one of greatest Muslim centers of worship in Cairo," said Mahmoud. "Most people in Egypt today are Muslims; a small number are Coptic Christians."

As he spoke, they heard the amplified call to worship from a nearby minaret and men began entering one door of the mosque.

"Why do some of the male worshippers have spots on their

foreheads?" asked Betsy Grover.

"That is from praying five times every day, as the Muslim faith dictates," said Mahmoud. "Usually men use prayer rugs, but sometimes the surfaces are very hard."

By mid-afternoon, Lisa was dragging her feet. The rumbling in her tummy had increased so she was sure other people could hear it, and she felt slightly nauseous.

"Hmm. Not the best day to get on an airplane," said James, referring to their upcoming flight to Aswan late that afternoon. "Maybe if I stuff you full of Imodium you'll make it. I wonder if short-hop planes in Egypt have bathrooms?"

Lisa glared at him. "Of course they do! Cairo and Aswan—and Luxor—have big planes as well as the smaller ones, and international airports. Egypt Air flies tourists all over this part of the globe. And modern Luxor is hardly a village in the middle of nowhere! It's the tourist capital of Upper Egypt."

"There, there. Keep your fur on. We'll see," he said, grinning maddeningly at her. "Think you'll make it through the carpet factory?"

"Maybe," growled Lisa. She was curious about the school, but she had a sneaking suspicion that the side trip, not on their itinerary, was a buying opportunity for the tourists and a commission opportunity for their guide.

First, they visited a large room where teenaged girls in long skirts and headscarves sat, nimble fingers flying as they knotted the wool. Then, Lisa was proved correct as they were herded to an upper level that was clearly the salesroom.

All around them were finished carpets of vibrantly colored wool and silk, stacked in neat piles, tacked onto the walls and central pillars, displayed on tables.

Immediately, a phalanx of smiling men in white shirts and brown or tan slacks appeared. They carried trays of glasses containing a bright red herbal tea and offered them to the Americans. Lisa hesitated, thinking maybe the tea was made from insufficiently boiled water (God forbid, maybe it came from that canal with the dead cow in it!), but then she shrugged and took one. Being a tourist was hard, thirsty work.

As soon as the tea was served, a smiling man descended upon each cluster of tourists, skillfully isolating couples and single travelers from each other.

Lisa and James were led to the smaller, more expensive silk rugs.

They were exquisite.

"See how it looks rough, then smooth," said their salesman, spinning a little carpet decorated with lotus blossoms on a table.

"Very pretty," said James, with a wink at Lisa. "But we are not interested in buying, only looking."

The salesman dismissed James as the less vulnerable buyer. "Madame, come this way, I have others to show you…" Lisa allowed herself to be led to a large stack of large rugs. She, too, could play this game. No one could force her to use a credit card if she didn't want to.

Behind her, she heard the ladies from Iowa cooing their enthusiasm. Judging from the amount of loot they'd already acquired, their Visa bills would be astounding.

She looked across the room. There was Mahmoud Sadaat, sitting relaxed with his buddies, sipping tea while his tourists earned him a nice little bonus.

Lisa, ignoring the salesman at her elbow, crossed over to talk with him. "Too bad Bertha couldn't join us today. I assume they put a cast on her arm? How is she feeling?"

Mahmoud's expression darkened. "Not good. The arm is okay, but the Doctor thinks she needs much rest for several days. She has fever and many aches."

"Oh, too bad!" said Lisa, whose head was now pounding.

She had a horrible feeling that Bertha wasn't going to be the only one getting sick in a foreign country.

Chapter Five

That evening, Cairo airport

It was after six p.m. After a long delay that included three metal detectors, James and Lisa and dozens of other tourists were milling around the waiting room at Cairo airport. While James strung his belt through the loops on his pants, stuffed his shirt back in, and put all the change and small objects back in his pockets, Lisa headed for the rest room and groaned when she saw the line of women snaking out the door.

She took a deep breath and said loudly, "Please, I feel really sick. I think I'm going to throw up." Lisa wasn't sure she was going to heave, but her gut was definitely rebelling and her need was urgent.

It worked. The line parted like the Red Sea in front of Moses and the Israelites, and she locked herself gratefully in a stall. As uncomfortable as she was inside, Lisa couldn't help hearing what two of the women were discussing.

"You know those tourists who got sick in Israel? About two weeks ago, I think it was."

"Yeah, they said it was some nasty virus—very contagious."

"Like, what are the symptoms?"

"Fever, rash, feeling crummy."

"Gee, I hope we don't get it here."

"We haven't been in Israel, silly…"

Lisa heard no more because nausea and intense cramps claimed her. After a grim interlude in the stall, Lisa mopped off her clammy face and washed her hands. Peering at herself in the mirror, she assessed her color now as alabaster pale rather than puke green.

Back in the waiting room, she found James with his head buried in an English newspaper. He ignored her, frowning at the printed page in front of him.

Lisa glanced at the article he was reading and zeroed in on the headline immediately: "Mysterious Disease at Megiddo..." She slumped into the chair next to him so she could read it too. Scanning the print, she saw that two French tourists had come down with high fevers, vomiting, and odd rashes. They were in a hospital in Jerusalem after visiting archaeological sites all over Israel. The Centers for Disease Control had been called in.

"James?"

"Yeah?"

"Isn't it unusual for the CDC to be called in?" She knew James had a special interest in the CDC because he'd almost become an Epidemiological Intelligence Officer to pay back loans for medical school.

"Not if the local doctors are having trouble identifying the illness," he said. "Could be a tropical disease that Israel isn't used to dealing with—something highly contagious."

Lisa tried to remember what little she knew about infectious diseases. "Most diseases have an incubation period, right? When a person is symptom free but capable of spreading virus or whatever it is?"

James nodded. "Too bad it's so easy to travel by airplane these days."

They looked at each other and Lisa said what they were both thinking.

"And Israel and Egypt aren't all that far apart." She shuddered and pulled her long-sleeved shirt closely around her.

The Nile undulated across the desert far below them.

Lisa pressed her nose against the cold glass of her airplane window.

Better than any textbook, the aerial view of the Nile revealed why Egyptians clustered around the great river. Vivid stripes of green flanked the river, while on both sides the beige of sand stretched endlessly. Life-giving water—the only major source for hundreds of miles, except for the odd oasis. Fertility, she thought. The key to settlement patterns the world over, but especially in Egypt and what was now Iraq, formerly called Mesopotamia, or the "Land between the Two Rivers."

Beside her, James shifted in his narrow seat. She knew he was uncomfortable; James always was on airplanes. Most economy class sections were not built with tall men in mind. James had to fold himself up like one of his son Sam's action figures in order to scrunch himself into a sitting position.

Lisa sighed with relief. Her stomach was quiet, for the moment, thanks to her good sense in marrying a doctor. James, always planning ahead, carried a little kit of miracle pills whenever he traveled abroad. He'd given her two Imodium—a wonderful, fast-acting drug. She was determined not to miss a single monument she'd come so far to see; she would just have to restrain her natural greed in restaurants and avoid the fresh stuff for the rest of the trip.

Lisa leaned her head against the seat and closed her eyes.

All too soon, the wing next to her seat dipped and the plane began its slow turn in preparation for landing. Lisa roused herself and looked at her husband. James was totally relaxed, his long frame spilling over the seat and both arm rests. His gentle mouth, partly disguised by his moustache, was slightly ajar and his eyes were closed. She could hear the gentle in and out of his breathing. Lisa smiled, thinking that the old saying "let sleeping babies lie" applied equally well to husbands.

The ground was closer now. She could see buildings, palm trees, and a tantalizing glimpse of temple. Was it the Temple of Luxor? Lisa felt the thudding of her heart as the fabled names of some of the greatest archaeological sites in the world scrolled across her mind: The Temple of Luxor. The Valley of the Kings. Karnak. The huge statues of Ramses II at Abu Simbel.

"Excited?" James asked, suddenly awake and grinning at her. His green eyes weren't sleepy at all.

"You know I am," she said, meeting his gaze and hugging herself. This was it—the best part of the trip, and the biggest cluster of monuments in all Egypt, was directly below them.

"I guess you'll have to take off your belt a few more times," she said.

James groaned as he tucked in his blue shirt and ran his long fingers through his bushy light brown hair.

The metal detectors were at the entrance of every archaeological site as well as at the Cairo Museum and every airport. And, as Lisa and James had already discovered, each busload of tourists was accompanied either by an armed guard riding inside the bus or a bunch of soldiers with AK's riding in a Toyota Land Rover nearby.

Their guide Mahmoud was strolling down the aisle. "Ladies and gentlemen," he said, beaming. "I apologize for delay."

Part of the delay had been Patsy Davis insisting on buying two more carpets after the rest of the group was already waiting in the bus.

"Tonight we sleep on Elephantine Island in very nice hotel. Tomorrow, at nine o'clock, we explore Aswan. Now, the Abu Simbel trip has been moved because we have problems with air reservations."

"Oh, God! Not another change in the itinerary!" said Ethel Bowman, who was seated a few rows behind them. Ethel liked everything cut-and-dried, with no changes in the pre-printed tour schedule. Mahmoud Sadaat's loosey-goosey style, with "extras" thrown in that made them late back to the hotel, was driving her crazy.

"Those of you who signed for that trip, we fly there after we do Nile Cruise from Aswan to Luxor. This is okay?" Mahmoud's olive skin gleamed with perspiration and earnestness.

Ethel and her husband grumbled, and everyone else began rooting around for hand luggage.

Chapter Six

Same Evening, En route to Aswan

"Cast not thy heart in pursuit of riches, for there is no ignoring Fate and Fortune."
(The Instruction of Amen-em-opet)

Zawi Farouk adjusted his seatbelt. He was a security guard for MISR Travel, currently on assignment with his buddy Mahmoud Sadaat. His job was to travel around with groups of Western tourists, coordinating the security arrangements with his Army buddies. It was a decent job, much better than working as a stonecutter, like his father. But not as good as his first job as a tour guide—too bad he'd been fired after the Luxor massacre when the Ministry of Tourism discovered his family ties to Al Gama'a Al Islamiyya—the militant group who had planned the massacre.

He groped for his cigarettes. Then he remembered he couldn't smoke until he got off the plane. Zawi's hand brushed the comforting bulk of his automatic pistol. He stretched his legs as far as they would go under the seat in front of him. Flights were getting as crowded and bumpy as mule-carts, a favorite mode of transportation around his hometown of Luxor.

The salary was inadequate, but the hours were reasonable. He had most of his evenings free to pursue other income to feed his large family. Zawi smiled grimly as he thought of his current project.

At least this group of American infidels was not too bad. Most of the tourists were biddable, meaning he and Mahmoud could herd them easily from monument to tourist trap factory (translation: buying opportunity) without fuss.

But he could have done without the Iowa ladies. Zawi frowned as he spied their ridiculous big hairdos three rows in front of him. No sense of proper dress at all—too much bare, wrinkled, and sun-damaged flesh. He shuddered delicately. In his family, the women were decently covered, though his sister-in-law, Safaa, liked gaudy colors in her headscarves instead of the regulation black. His own sweet Mina was much more discreet, always wearing the long, black abiyah and correct dark headscarf. She had managed to bring up at least two of their three daughters as modest, Allah-fearing Muslims.

Zawi wished his father were still alive. He had been an alabaster worker before his final accident. A stern, tough father who hid his affections most of the time and managed his family with an iron hand. Despite his tendency to yell first and explain later, Zawi had never doubted that his father loved him. He wondered how his father would have handled Zawi's youngest daughter, Zara, who was turning out to be quite a handful. She had a mind of her own, that one, and was resistant to traditional rules and expectations. She wanted more education, but for what? Even a college degree was no guarantee of a job, these days. And these women who worked—they started imitating Western habits and tastes, growing away from their families… Time to marry her off, preferably to someone older. Get her away from that no-good youth, Farzad Hammadi, who had no prospects. Perhaps Achmed? Zawi smiled. That would be the best way to water two camels at the same oasis—marrying off his daughter while cementing a business alliance.

He frowned as he recalled the conversation he'd had at his favorite ahwa, or coffeehouse, only a week ago. Zawi and his brother Ali and Mustafa took turns with the water pipe, enjoying the fragrance of the tobacco as they sipped their coriander-flavored coffee.

"Times are changing," said Ali, taking a leisurely puff. "When Safaa and I married, there was no question about roles for men and women. Men were heads of households and sole wage earners—you could support a family on one salary in those days—and wives and daughters took care of the cooking and childcare. Islam was taken for granted; it was integrated into every aspect of life. We were all happier in those days."

Mustafa grunted. "Tradition has gone out the window," he said. "My youngest son, he wants to move to Cairo. Can you imagine! I will never let him go, even if I could afford to send him to our cousin there." Mustafa had never been to Cairo; Zawi was the only one of the three brothers who had been that far north. "He would come under evil influences—American music, alcohol, loose women—it does not bear thinking about."

"Yes, but he would be able to earn more money, as I did right after I got my tourism degree."

"Oh, and that degree did you so much good! You lost the guide job so now you rely on your army friends!" scoffed Mustafa.

"It's not my fault I lost that job," growled Zawi. "We were talking about Cairo. My son Abdul wants to be a doctor and to start his training immediately. The best training programs are in Cairo, and I cannot pay for six children here in Luxor and support one in the big city, even if he did live with Rashid. There are registration fees, books, and I would have to contribute for his food. I want him to wait a year so I can replace the income he brings me, but he is so impatient! And my daughter too has ideas about study in Cairo. She is good with numbers."

"They would both grow away from the faith if you sent them there," said Mustafa, scowling into his coffee. "Even if they lived in Rashid's household. He is less observant; he does not think as we do. We must keep our children close to home."

"It's like the mummy's curse," said Ali. " 'He who disturbs this royal tomb shall surely die,' only for us it's 'He who lets Western ways into his home shall surely suffer.' "

"Westerners!" scoffed Mustafa. "They have no business living and working in our country."

"But Western tourists are our livelihood," said Zawi, taking another puff.

"Americans are the worst!" said Mustafa, falling into his usual Jihadic rant. "They support the Israelis—damn their Jewish hearts."

Zawi returned to his perusal of the Western tourists he could see from his airplane seat. There was that blond woman, Li-something Donahue, five rows ahead. She was all right. She kept her fair skin covered with long-sleeved shirts and long pants. That long, light hair and those sensuous lips—oooh la la. She was something any sheik would prize in his harem.

Zawi closed his eyes, switching gears to the next phase in the Farouk brothers' current enterprise. He wished he could be sure Ali had a grip on all the details of the transfer. There was too much danger of leaks, and his old Army unit was watching him. They knew too much about his early career as an antiquities smuggler, before the laws tightened up and the threat of imprisonment and heavy fines forced him to look for a more conventional job. Zawi had to strike the right balance between doing favors for his Army buddies and feathering his own nest.

His lips twisted as he considered his elder brother's incompetence. The operation would run so much smoother if he, Zawi, were in charge. The next package would arrive from Cairo in a day or two. He and his brothers would have a week to finalize the details of distribution, and then they would strike. But he had yet to work out the mechanism of transmission.

The Hamas contact in Israel hadn't been able to share much information about the experiment at Megiddo. And where, exactly, did the funds for all this originate? He thought it was Hezbollah in Syria, but it could easily be Iran. This work was so compartmentalized; it made Zawi nervous to have so little control over the operation.

The details of the sickness and how it spread, that was the key to success. His team must decide whether to spray tourists in a crowded souk with a solution of the powder or to try smearing it on white shirts and those ridiculous belly dance costumes hanging above every stall. Before he made that decision, Zawi wanted to ask someone what would be most effective.

He shifted in his seat as he mentally ran down a list of names.

Ah. Mustafa, his brother, had a friend in Aswan who had medical training. He was the one.

The Luxor market was teeming with people and animals. Women shopped for their daily meals, lugging bulging string bags, while children raced from booth to booth. Dogs scavenged under tables and tripped unwary shoppers. The best customers were the stray cats, who snuck into narrow spaces with purloined fishheads or scraps of meat.

Mina Farouk had finished selling her goat cheese. Now she was on automatic pilot, bargaining for fresh lamb and the choicest vegetables while her mind was engaged elsewhere.

She was worried about her daughter, Zara. Zara was her youngest girl, her brightest and most beautiful. And Mina was sure her husband Zawi was beginning to plan for an auspicious marriage that would benefit his bank account at the expense of family harmony. Zawi was a hard man at the best of times, but on such an issue he would be implacable.

Mina was the middle child of six from a carpet manufacturing family in Aswan. Although she had barely more than an eighth grade education, she had a fierce belief in women's rights in a changing world. Her father had married her off at fifteen, selling her into a marriage she didn't want and was in no way prepared for. Her body was old at thirty-eight after years of child-bearing and variable nutrition.

Her twenty-year-old twins, Alia and Abeeta, had finished two years of high school but had also married young. They were now burdened with small children and could not work outside the home. Her sons Mohammed, Ali, and Abdul (twenty-two, twenty-one, and seventeen) sold canvas bags and other trinkets to tourists. None of the older sons could afford to move their growing families away from the main house.

Mina wanted better for her youngest daughter. Zara was at her peak of loveliness and marketability. She was only sixteen, but with a sharp mind and a hunger for more learning. Maybe she really could go to university, as her teacher said. Tuition was free, but how would they find money for her registration fees, books, clothes, and food? And if she went to Zawi's cousin, would anyone really look after her?

Mina shook her head and placed three cucumbers in her bag. She headed for the baker's stall. As she approached the stacked pita breads and long loaves, a mangy white cat leapt down from the awning, captured a pita, and sped away down the alley.

Good for it, thought Mina, mentally cheering for all poor and downtrodden creatures. Snatch at the good things in life when you can because you may never have another chance.

"Six pitas, please, and make sure they are fresh," she said to the vendor as she handed over some small coins as baksheesh. "And I'll take four of the almond pastries."

Nodding, the woman wrapped the pitas in paper and handed them over. "Your daughter is not with you today?" she asked.

"No, Zara is minding the little ones—her nieces and nephews—at home," answered Mina. If Zara was married off, she and Alia would have even less help around the house. No matter, it was Zara's happiness that mattered most.

She trudged down the alley, ignoring the shouts of "fresh tilapia!" and "buy your pastries here, hot out of the oven!" Her strategy so far in her marriage had been to let Zawi have his way in most decisions as long as they didn't affect the children. That way, she preserved some peace in the household although usually at her own expense. This time, however, they were headed for a major confrontation.

Zara would never marry anyone willingly unless it was Farzad Hammadi. Mina had watched their relationship develop and was satisfied that it was the real thing, genuine and lasting affection, mixed with something she had rarely seen in a couple so young: mutual respect. Farzad knew Zara was smart, smarter than he was at numbers. He encouraged her to pursue accounting training, saying he would make her a business partner when she could keep the books. A partnership. A marriage based on true love and equality.

Mina shifted the heavy string bag from one shoulder to the other. Zawi would insist on an older man, a business partner who would chip into Zara's dowry so Zawi wouldn't have to pay so much for the wedding. So, if she couldn't sway his opinion, what could she do?

Help the young lovers elope, with enough money to start on and a safe place to stay until they found their own little house.

She lifted her hijab away from her sweating face and began to plan.

Chapter Seven

Tuesday-Thursday, September 7-9

Aswan to Luxor (Nile Cruise).

The next few days passed in a haze of cramps and fatigue as King Tut's Revenge (or was it the Ramses Runs?) caught up with Lisa. Years later, she recalled vivid vignettes of the monuments she'd come so far to see—lotus columns, Horus falcons, hieroglyphs, the crocodile god—interspersed with embarrassing episodes on white porcelain thrones in unsalubrious locations. To add insult to injury, the coarse Egyptian toilet paper resembled crushed newspaper, and there was never enough of it.

It was after nine at night when they finally arrived at the Oberoi Hotel Aswan.

As they docked at the hotel quay on Elephantine Island in the middle of a wide section of the Nile, Lisa gaped at the extraordinary concrete edifice in front of her. It boasted a glass and steel superstructure that resembled nothing so much as an airport control tower.

Everyone except Lisa was ravenously hungry after the full day in Cairo and a session in Purgatory (being trapped in Cairo airport security at its busiest) and then the early evening flight. When Lisa saw the crisp linens and four, beautiful plump pillows in their room, she told James to go to dinner without her. After swallowing two glasses of bottled water and another Imodium, she collapsed on the soft mattress without even changing to her pajamas.

Breakfast the next morning was served buffet style. Platters of salads jostled with hot trays full of scrambled eggs, peculiar-looking sausage, and a starchy item that Lisa knew was not potato. Shuddering at the abundance of greasy and sugared foods she dared not eat, Lisa chose a single, hard-boiled egg, a dry roll, and a scalding cup of tea.

Betsy Grover, while consuming a substantial plate of goodies and black coffee, watched her eat this meager breakfast. "Won't you starve, doing a full day of tourism on that little food?"

"I'll manage. You don't know how sick I've been," replied Lisa, sipping her tea. At least she didn't have a fever—James had been checking her twice a day.

James joined them, bearing a full plate of eggs, sausage, pastries, and fresh mango slices. He seated himself next to a Korean couple across from Betsy and produced a small package with a flourish. "Lisa, this is what the guide ordered for you." He handed her some pills and pointed to a flimsy blue paper with writing in Arabic and English.

Lisa's head hurt too much for reading. "Just tell me what it says."

He examined it. "Wait, it's some prescription for amoebic dysentery! I know you're not that bad—yet. But maybe I'll save it as a souvenir." He took the pills back from her and stuck them in his pocket. "Hey, Betsy, did you ever see cholera?"

Lisa listened cynically as James and Betsy grew animated. It was hard to believe people could flirt while exchanging the gory details of death and disease, but these two managed. Just like medical students swapping stories over their first anatomy corpse.

After breakfast, the tour group boarded the bus to tour around Aswan. The first stop was the granite quarries where they gaped at the Unfinished Obelisk—an enormous monument that had been three-quarters completed before the stone cracked and the workmen abandoned it.

At Aswan, Mahmoud told them how the two dams—the northern dam and the southern High Dam, had been built in stages over almost forty years, and how they'd changed the economy of Nubia and southern Egypt.

"We used to have crocodiles here; now they are all in Lake Nasser, behind the High Dam. Many people had to move when the water started to rise, and farmers lost their traditional crop of dates. And the moving of the monuments—a huge project, very expensive. In two days, some of you see Abu Simbel, the most impressive temple, moved block by block. But I show you also Philae, moved from an island to the mainland. And one other temple, the Temple of Dendur, was moved from here to New York. It is intact in your Metropolitan Museum."

There was an appreciative murmur as everyone thought about the amount of effort and money to complete the moves.

"Mahmoud, didn't you say the Russians helped with the High

Dam?" asked Betsy, who was standing, as usual, next to James.

"That is correct, Ms. Grover. You will see the Lotus Monument that remembers this international cooperation."

Later, on the way to board their cruise ship, Mahmoud looked at his watch and decided on another abrupt change of plan. He took them to a souk that sold gold and silver ankhs, the Egyptian symbol of eternal life. The Iowa ladies bought several for themselves and friends and relatives back home.

Finally, the bus rumbled onto the long quay where the Oberoi hotel boats docked. Much smaller than the sort of American cruise ships that crisscrossed the Caribbean, these boats were long and low to the water. Each one had four levels, with curtained passenger rooms on the lower levels and a passenger deck with a small swimming pool and potted plants on its roof. One boat could hold eighty-ninety passengers—about two busloads—plus crew.

James, with Lisa leaning tiredly against him, walked down a ramp to a white boat with blue trim. They passed through a metal detector that James' belt set off. He paused, but none of the boat staff seem concerned, so he kept moving. This boat's lobby was quite large, with thick carpets and gold-flecked wallpaper and an impressive spiral staircase.

Lisa and James stepped over another, narrower ramp into their own boat, the Shehrayar. They and the other tourists passed through a second metal detector (again, no one paid any attention to the numerous beeps) and found themselves in a smaller, less elegant lobby with a front desk, a few upholstered chairs, and a straight, utilitarian staircase.

"Why is our boat moored next to the bigger one?" asked Betsy Grover.

Mahmoud told them this was common at the height of the tourist season, to stack the cruise boats one against the other because the quay wasn't long enough to accommodate all of them.

"Now, everybody," said Mahmoud, raising his voice so he could be overheard in the lobby. "You have half an hour to freshen up, then meet back here for felucca sail. "Dinner will be later than planned, but I promise you it will be worth it," he said.

Tired as she was, Lisa was pleased to see that their room was on the outside of the two joined boats, so they had a Nile view. Too bad for the tourists on the other side of the boat—they had no view except for someone else's closed curtains.

In fact, she and James had a double picture window. Lisa looked around her and realized that everything was doubled: older, smaller rooms had been thrown together to make the kind of staterooms American tourists expected: wide beds (two twins sheeted together) and the extra benefit of two miniscule bathrooms instead of only one.

Lisa stepped up to the window. The water was murky, and they could see feathery-topped plants along the shore. "Papyrus?" she wondered out loud.

"And water hyacinth, just like the stuff in my mother's pond," said James, pointing to dark green plants bobbing in the current.

Lisa looked longingly at the bed before washing her face, swallowing yet another Imodium, and joining the others.

The felucca, a traditional Egyptian sailboat, boasted a single trapezoidal sail, much patched. The two sailors wore long white robes that covered them from neck to ankle.

The sun was setting as they pulled away from the dock, revealing stark silhouettes of minarets on the far shore. Lisa slumped on a hard bench between James and Betsy. Then the cooler air on the water revived her. She sat up and observed her tour mates, who were relaxed and laughing at the antics of the sailors.

"Now, we sing Nubian song!" said the older sailor. He wore a long, white gelabiyah but no turban or scarf. He pulled out a shallow drum and began beating it, singing simple syllables in a regular rhythm they were all able to imitate.

Soon the tourists were clapping and swaying and Mahmoud's eyes were twinkling with satisfaction. Lisa guessed what he was thinking: happy tourists meant bigger tips when they finally returned to Cairo.

The song finished, and the younger sailor motioned to a couple of women to move off the central bench. He pulled off the canvas cover, and surprise, surprise, there were piles of necklaces and bracelets, all for sale! How clever, thought Lisa, to wait until they were in the middle of the Nile. Talk about a captive audience!

She looked at James, who was smiling. "Don't buy from a Nubian in the dark," he whispered. Lisa giggled as she realized just how dark it was—too dark to see the color of the beads in the necklaces that were surely plastic and not granite from the Aswan quarry as the sailors turned salesmen claimed. The wares showed up well against the white gelabiyahs of the sailors, but not as well as the gleam of their white teeth as they smilingly accepted dozens of dollar bills.

Overnight the cruise ship sailed north to Edfu. The next morning, Lisa felt a little better. She ate toast with her boiled egg and joined James for the morning tour of the Temple of Horus via an early morning horse-and-buggy ride.

As they left the boat, a crew member handed each tourist a plastic token, requesting that each of them return the token as they re-boarded the boat at lunch time. Lisa reflected that this was a primitive but effective way of counting heads before they cruised to their next stop along the Nile.

She climbed the steps to street level. At the top, Lisa was helped into a rickety cart with a swaying basket chair and no railings. As the horse began clip-clopping at a speed that threatened to tip them both into the street, she watched the metal fender over her wheel quiver over a frayed tire that had no discernable tread.

"James!" she hissed, grabbing her husband for support. "I don't think this cart is road worthy!"

"I agree," he said, with a rather worried smile. "I think it probably belongs in a museum."

Their young driver turned around and flashed a grin. "You like? Go faster?"

"No! I mean, this is fine," said Lisa, clutching the leather seat as it sagged towards the pavement.

She distracted herself by looking at the surrounding neighborhood. The poverty was palpable. Houses were mere shacks of mudbrick, roofed with corrugated tin. Clusters of young girls wearing navy uniforms and white headscarves congregated on the way to school next to abandoned lots full of garbage. Two boys, not bound for school, scaled a mountain of refuse carrying plastic bags. What were they scavenging for—metal scraps, bottles, plastic? They trundled past other vehicles—Fiats, bicycles, and ancient trucks bearing gas cans and crates full of chickens.

They passed an overturned cart, fortunately with no one inside it. Lisa and James looked at each other after they noticed that one wheel had actually fallen off.

"It figures," said James.

After about twenty nail-biting minutes, they arrived at the temple parking lot, and their driver pulled up with a final burst of speed.

"Me, Mohammed!" he said, pointing to his chest, when Lisa gave him his tip. "You American?"

"No, Swedish," replied Lisa, who was feeling ornery.

The Edfu temple had an imposing entrance guarded by a falcon, the bird incarnation of the sky god, Horus. Lisa and her group gathered around Mahmoud to hear his introduction.

"This temple is one of the best preserved under Greek rule," he began. "But you will notice bullet holes and sword damage from Napoleon's visit here in early 1800s. His soldiers like to make sport," he pointed at a sun disk over the entrance, "by shooting up there. And they sharpened their blades over here." Lisa saw the shallow grooves in the soft stone that obscured the fine carving underneath and shook her head. Every archaeological site had problems with vandalism, but she'd never seen this particular form before.

"And here is Passage of Victory," continued Mahmoud. "It is the triumph of Horus over Seth." Another Cain and Abel story, thought Lisa; two brothers fighting for a kingdom, only the Egyptian brothers were both gods.

Once again, Mahmoud Sadaat diverted the group from their printed schedule, this time to a perfume factory. Again, they were served tea and herded into smaller groups by slick salesmen. This stop Lisa actually enjoyed; she volunteered to try the headache cure, "all free, no obligation, Madame." It turned out to be a vigorous head massage. Lisa was led to a rose-colored couch some distance away from the other tourists, where strong fingers anointed her temples with fragrant sandalwood oil. She closed her eyes, giving herself up to the soothing sensations. When she opened her eyes, her hair felt greasy and stood out from her head in clumps, but the headache was gone.

"Do I need to pay you?" she asked the young man who had performed the miracle.

"Not necessary," he said, smiling, and moved away.

Walking over to a tray of essence samples—frankincense and myrrh, among others—she saw that the other tourists were enthusiastically buying collections of perfume in delicate glass bottles. The Iowa ladies were in full buying mode, stuffing their daypacks with presents. No wonder her masseur hadn't felt the need of a tip; he and his colleagues, as well as Mahmoud, would clean up from this visit.

The next day, Lisa dragged herself up to the top deck with the others to watch the cruise ship go through the locks at Esna. This was often a long business depending on how many cruise boats were on that section of the Nile at the same time.

"I've never been through a lock before," said Betsy Grover eagerly. She sidled over to stand next to James.

Lisa was getting a little tired of Betsy's obvious interest in James. Maybe, she thought, trying to excuse the other woman, Betsy's long experience as a solo traveler made her latch onto interesting people so she wouldn't have to be alone. But James was enjoying the attention a little too much.

The boat carried another tour group along with Lisa's, including a mixture of Australian, Indian, and British travelers. Two or three people frolicked in the postage-stamp sized swimming pool. Everyone else lounged around between the potted plants, leaning over the railing, chatting and pointing as the level of the water rose after the boat moved into the first compartment of the lock.

Lisa wandered over to join Nancy at the rail. "How's Bertha? Have you been able to call her hospital from the cruise boat?"

Nancy frowned and shifted her enormous handbag to the other shoulder. "She still has a fever, and now a rash. The doctor thinks it's shingles."

Lisa nodded. "From the chicken pox virus. It lives on in your body after you get the childhood disease."

"Oh, I don't know about that. My aunt had it, and she had an awful time."

"Poor Bertha," said Lisa. "Shingles on top of a broken arm and missing the cruise."

Nancy shook her head dismally.

"Look!" said David Wallis, the Chicago dentist. "What is that guy doing?" He pointed to a young man in a dilapidated rowboat on the near side of the lock who was slapping the water with a stick.

"I think he's attracting fish that way," said James. "See the net in the back of his skiff? I saw him scoop up something with that a few minutes ago."

On the shore, a few yards beyond the fishing boat, was a cattle shed built of rushes—a crude shelter for the animals so they could spend a little time out of the hot sun. Lisa spotted white ibises in an adjoining canal. Such glimpses of another lifestyle made her feel she was no longer in the twenty-first century.

All too soon, however, her insides rumbled in warning. Lisa hustled down the stairway to her stateroom and private bathroom.

She flushed the toilet and lay on the bed, mesmerized by the novel view. Outside the window, the wooden walls of the lock appeared to move past her. Gradually, the boat sank down to the lock's lowest level, leaving the sunlight behind.

A few hours later, they docked at Luxor. This stop marked the end of the Nile Cruise, but Lisa remembered that they had two more nights in the Shehrayar. Everyone was happy with this arrangement since the food was so good on the boat. Besides, it was bliss not to have to pack and unpack every day.

Lisa realized that her stomach was quiet. Maybe she could even enjoy dinner, consuming something a little more substantial than a boiled egg. And the best part of the trip was yet to come—the Valley of the Kings and Queens, and Karnak. She would finally see the greatest temple complex in all Egypt and walk on the Avenue of the Sphinxes.

James picked up an English newspaper at the front desk on the way to their room.

"Uh oh," he said scanning the headlines.

"What?" said Lisa, struggling to maneuver around the Iowa ladies, who had halted in a clump near the staircase.

"There are more cases of that odd sickness—one in Alexandria, in northern Egypt," he said in a low voice. "And one at Petra in Jordan. Whatever it is, it's spreading."

Chapter Eight

Friday morning, September 10, Luxor

"Go down, Moses, way down in Egypt's land. Tell ol' Pharaoh to let my people go."
(American spiritual)

After another night spent on the luxurious hotel boat, the tour group arrived at the fabled site of Karnak. This massive complex dedicated to the god Amun, located on the East Bank of the Nile, was a "must see" for all visitors to Egypt.

"This was major center worship in Eighteenth Dynasty and before," Mahmoud told them. "And in ancient Egyptian language, Karnak was called 'The Most Perfect of Places.' Original sanctuary was built during the Middle Kingdom Period, but almost every Pharaoh built something here so you will see many architecture styles stretching over more than thirteen hundred years. Karnak was dedicated to the ram-headed god Amun. You see him in the sphinxes on the short avenue here on the south side."

James, a little bored with Mahmoud's lecturing, wandered away from the group. He leaned over to take a close-up photograph of a ram-headed sphinx. He had promised Lisa to take backups for her museum and sphinx articles in case her own digital camera failed in some way. They both had a healthy distrust of technology—computers failed all the time, especially when you needed more memory or a timely printout. There was no such thing as too many backups. He pulled out a white cotton hanky and wiped his sweaty forehead.

Phew, must be over a hundred easily. And the airborne sand was everywhere; it crept into his ears and nostrils every time he moved.

Lisa was right. This place was amazing. They'd watched several videos at home before they'd come, but Lisa had said no footage could convey the true size of the pillars of the hypostyle hall at Karnak, or the grandeur of the Avenue of Sphinxes—this time, human-headed ones—that stretched from the temple complex of Karnak all the way to the Temple of Luxor so ancient worshippers could traverse the distance easily. She'd quoted the nineteenth century writer Amelia Edwards: "It is a place that has been much written about...the scale is too vast; the effect too tremendous; the sense of one's own dumbness and littleness, and incapacity, too complete and crushing."

He clicked the button on his Olympus camera that reviewed pictures already taken. Yup, he'd gotten some great angled views of those pillars with tourists to show the scale. Unbelievably huge—and with a surprising amount of colored pigment remaining on the carved reliefs.

"Ram's heads, all the way," said his wife at his elbow. James turned to watch Lisa as she paced out the width of the avenue between the Precinct of Amun and the Precinct of Mut. "Sacred to the god Amun, part of the Theban triad."

He smiled at her furrowed brow with the dark gold bangs peeking out from under her broad-brimmed sunhat. James found his wife beautiful, although some might say Lisa's nose and chin were too prominent. The changing expressions on her mobile face gave her beauty distinction far more than her blond hair or slender figure. His gaze moved to her belly, gently rounded now that she was in her mid-thirties. She looked good in those khaki pants.

Lisa stopped staring at sphinxes and asked, "Did you like the Sacred Lake?"

"It was okay. But what I really liked was the giant scarab beetle."

James was referring to the stone statue that tour guides told their charges to walk around for good luck—three times for a marriage, seven times for a first child. Since he and Lisa already had enough children, they'd only walked around the scarab three times. James thought they had an excellent marriage, but what the hell, a little more good fortune couldn't hurt.

Lisa glanced at her watch. "We've still got time," she said. "The La-La ladies from Iowa will be in the souk for at least twenty minutes. Let's walk the avenue south toward Luxor."

"Okay."

Together they examined the crumbling remnants of human-headed sphinx statues that lurked like doorstops in the front yards of the modern dwellings. These shack-like buildings—as usual, constructed mostly of mudbrick with corrugated tin roofs—clustered along sections of the ancient avenue.

James was busily taking photos when he heard feminine shrieking.

He turned around and saw a knot of women in black abiyahs and headscarves gesticulating. One was hunched over, crying. Then a tall youth yelled at the milling tourists around James. "Doctor-e? Doctor-e? Anyone doctor-e here?"

"I am a doctor," James said, with some reluctance. His position here in Egypt was dubious, to say the least. James could not practice medicine legally outside the United States. Also, he was on holiday. But what if he got involved in something major, without meaning to? Surely...

"Please, you come," said the youth, tugging at his sleeve. "You come right now." James looked for his wife and found her sauntering past a Moslem graveyard with domed monuments at least two blocks away. He caught her attention by waving and signaled that he was going inside.

Realizing his moment of decision was already past, James followed the Egyptian youth into a small, cramped house whose construction appeared to combine both cheap modern brick and ancient, sun-dried mudbrick. There were hardly any windows and no ceiling fan or air-conditioning. It must be a hundred degrees in here, and a long, long way from a sterile, Western radiology suite.

James stopped in the narrow hallway. He found himself in semi-darkness.

"This way. My aunt, she very sick," said the boy, who appeared to be about fifteen.

"Where is her husband?" asked James, thinking of the likely scandal if he examined a Moslem woman without her husband present.

"He at work. You come now."

Shrugging, James entered a small, crowded bedroom. He could barely see because the small window was covered by a piece of sacking.

He pushed aside the makeshift curtain and was dazzled by sunlight pouring in. "Where is the patient?" he asked, hoping the boy understood his few words of Arabic.

"Right here, on the bed," said the boy.

James' eyes adjusted, and he saw a middle-aged, coffee-skinned Arab with straggly brown hair and huge dark eyes slumped on a double bed. Judging by the number of shoes and clothes strewn on the floor, it probably served as a bedroom for several other people at night.

He stepped closer. Now he could see that the woman's face was suffused with a peculiar, bumpy red rash. Several large pustules spotted her cheeks and chin.

James stared, memorizing the details while the sweat poured down his neck and back. The lesions were well circumscribed and round, each with a depressed area, like a belly button. Like pictures he'd seen in med school, but never in the flesh.

Swallowing with difficulty, James gently pulled down the thin cotton coverlet and examined her chest and hands. More pustules.

He'd never actually seen a case of smallpox. No one of his generation had. The last confirmed case was nearly thirty years ago. Frantically James tried to remember the details. How was the disease transmitted? What was the incubation period?

"What is your name?" he asked the boy.

"Ali Farouk, the younger one. I am called after my uncle."

"Wait here. Don't let anyone else in the house but me. I am going to find another doctor. Okay?"

"Is okay."

James backed out of the room and hustled to the front door. Outside, Lisa stood looking at something to the right of the door.

He gulped greedily at the cooler air and grabbed her arm. "Lisa, I really need help. Medical emergency. Go see if you can find Stanley Rosenberg. And get Mahmoud Sadaat. Ask him to find a member of the Tourist Police. Tell him we need help keeping relatives and friends away from this house. But don't attract unnecessary attention; don't tell any of the other tourists."

She turned wide blue eyes on him. "You look terrible. What's wrong in there?"

"We've got—we've got a bad situation. There's a woman who's very sick in there. She needs immediate help." Not to mention immediate isolation, he added to himself. "Go, please—and hurry."

James watched his wife stride down the tourist path and felt the sweat turn cold on his back. Had he been exposed, or not? He wished he could remember if smallpox could be transmitted through a cough or a sneeze. Had the sick woman coughed in his face? Oh, God. He'd touched her garment, not her skin, right? James hadn't been so scared since he was twenty, facing an operation on a benign tumor that was in a tricky location on his throat.

The retired New York physician, Stanley Rosenberg, appeared in minutes. James heaved a sigh of relief.

"Lisa said you needed help. Don't tell me we have another broken arm?" asked Stanley.

"No," said James. "It's much worse than that." He hoped he wasn't shaking.

"Are you sure we should get involved? I mean, we're foreigners here…"

"Too late—I am involved. You have to see this case," said James with a groan.

Stanley seemed to grow taller. "Show me," he said.

Chapter Nine

Same morning, Luxor

"Take handfuls of soot from a furnace and have Moses toss it into the air in the presence of Pharaoh. It will become fine dust over the whole land of Egypt, and festering boils will break out on men." (Exodus 9:8-9)

Lisa's heart slammed in her chest as she looked around for Mahmoud or the tourist police. James' behavior had alarmed her. What wasn't he telling her? She pushed past the teeming tourists, resisting the tugging on her long sleeves by youths trying to sell her brass trinkets and canvas bags. Her Arabic was inadequate for this situation, but she drew in a deep breath and then yelled, "Police! Police!" several times, hoping one of them would understand.

One of the smaller boys dashed away. Lisa couldn't tell if he was trying to be helpful, or just chasing after the next group of tourists heading for the temple complex.

Miraculously, Stanley appeared. After she sent him over to join James, she spotted Mahmoud Sadaat near the beginning of the Avenue of the Sphinxes and sprinted the remaining distance between them.

"Mahmoud!" Lisa gasped. "There's a very sick woman in a house back there. Can you use your cell phone to call an ambulance? Tell them my husband—he's a doctor—says she needs immediate medical attention. Dr. Rosenberg is in there, too. "

Mahmoud's pale olive face tightened as he pulled out his phone.

Lisa put a hand on his arm. "And we need Tourist Police. To keep everyone—all the friends and relatives—away from that house."

He nodded, beginning to sweat. "Our group is finished here, anyway. And the other tour groups can stay away from this part of Avenue of Sphinxes."

Mahmoud finished his calls and moved to collect the tourists in Lisa's group. "Okay, time to go! Time for lunch. Back to the bus!" His forehead glistened.

Lisa could well imagine his dismay. It was the second medical incident he had encountered on one tour. It might not directly impact his group, but it was hardly good publicity for MISR Travel and other tour companies like it.

Lisa thought about going back to help James and Stanley. But James had been jittery with tension. Whatever it was, he didn't need her. Not for the first time, Lisa wished she had medical training so she wouldn't feel so helpless. Shrugging off her frustration, Lisa followed the others back toward the bus. Everyone seemed to know already there was a medical emergency.

"What's going on, Lisa?" said Patsy Davis. "Did you see anything?"

"No," said Lisa accurately. "But my husband is a doctor and he's helping. So is Stanley Rosenberg. I'm sure they'll tell us when they come back."

"Doctors never tell me anything," said Patsy with a sniff.

Lisa suppressed a smile. If she were a doctor, she'd keep quiet around Patsy Davis, too.

"Was it a heart attack?" asked David Wallis, the heavy fifty-five-year-old dentist from Chicago. He looked like a poster child for heart disease himself—his round, red face showed the telltale signs of prolonged, liberal consumption of rich food and alcohol.

"I don't know."

"Which house is it?" asked Betsy Grover. "Maybe I can help."

"It's that tan one in the middle. It's mudbrick, but it looks stuccoed," said Lisa. "But I don't think they need more people..." Her voice trailed off as she watched Betsy walk away up the Avenue of the Sphinxes.

Lisa reached the gaping doors of the air-conditioned bus and climbed gratefully aboard. The tour group's driver, a forty-something, olive-skinned man named Issam, smiled at her as she paid him in American one-dollar bills for another two bottles of sealed water. That fistful of ones she'd grabbed right before they'd flown to Egypt was proving mighty useful. The vendors clearly preferred American cash—especially crisp, new bills—that they could sit on, waiting for the best exchange rate and thus increasing their meager profits.

The rest of the group straggled onto the bus, clutching guidebooks, cameras, and souvenirs.

Lisa sat down in her seat with a thump and unscrewed one of the bottle caps. She moistened her handkerchief with some of the precious water and folded the damp cloth over her hot forehead. When she opened her eyes, she saw James climbing onto the bus.

All eyes turned his way.

"Hey, Barber! What happened?"

"Is someone hurt?"

James swung into the seat and twisted his head around to field questions. "An Egyptian woman's pretty sick. Some kind of intestinal flu, I think. I got called in because there was no other doctor nearby."

He turned back towards Lisa and met her gaze. His face was bland and expressionless.

She knew he was lying. She'd have to get him alone and find out what the hell was going on.

Lisa handed James a cold bottle of water while she plotted how to get him off to one side during lunch.

"Thanks," he said, unscrewing the cap and downing half of it. "I needed that."

He rubbed his eyes wearily.

Stanley and Betsy climbed on and the bus started moving.

Mahmoud Sadaat picked up the bus microphone. "Ladies and gentlemen, I have some news," he began.

"Does it involve lunch?" cried a woman at the back of the bus.

"Did someone have a heart attack?" asked Wallis.

"Yes, ma'am, all in good time. No, sir, I do not know what the medical problem is. Now, we are going back to our very nice hotel boat. We have our lunch there. After lunch, we will decide how to spend the afternoon."

"But why aren't we sticking to the printed schedule?" cried Patsy Davis. "I'm tired of all these changes and delays."

Other tourists joined in, and a surly muttering rose like a cloud of dust.

Poor Mahmoud. Once again he had to deal with a bunch of people who were used to instant service and continual gratification. Lisa decided he couldn't be earning enough to compensate for all the aggravation he experienced. Early in her career as an archaeologist, she'd envied colleagues who'd been invited to lead tours to Greece or the Near East, all expenses paid. No more. She had no desire to be a professional tour guide—it was too much hard work.

But for the moment, she was much more worried about the sick woman James had seen—and what that might mean.

"Okay, now tell me. What's going on?" asked Lisa, once they were safely in their double stateroom on the Shehrayar. She dumped her daypack on the queen-sized bed.

James faced her. The muscles at the back of his neck tightened. "I may have been exposed to a particularly nasty disease. I don't want you to get it."

Lisa stood stock still, staring at him as her face paled.

"What disease?" she whispered.

He wished he didn't have to frighten her. "I think that woman back there on the Avenue of the Sphinxes has smallpox—and that's a medical emergency because it's contagious and there's no cure. Stan Rosenberg was there with me after you left. He thinks I'm right—and that I should not sleep in the same bed with you until we know for sure."

"Holy Toledo," said Lisa, exhaling loudly and plunking herself down on one side of the bed. "How do you find out for sure?"

James unbuttoned his filthy, sweat-stained shirt and tore it off as if it were on fire. "I don't know the system over here. But the World Health Organization has been called. Stan says first they'll do saliva and blood tests to see if it's really smallpox. Then, the CDC in Atlanta, and maybe NAMRU will be called in. If smallpox is confirmed, they'll try to quarantine that house on the Avenue of the Sphinxes—almost impossible, I imagine, with the number of people who appear to live there." He rubbed his hand over his face. Some vacation from medicine this was turning out to be.

"What's NAMRU?"

"It's a U.S. Navy outfit—started during World War II. I think NAMRU stands for 'Navy Medical Research Unit,' or something like that. They did early work on cholera and typhus and started a lab for other infectious diseases in the Cairo area. They're connected with WHO, too."

"How contagious is it, exactly, if it is smallpox?"

"If I'm remembering right, it spreads through the fluids that form into pustules and scabs that are shed from the skin. I don't think you can get it through the air, by someone sneezing in your face like with the common cold. But on the other hand…" He frowned and looked away from her. James didn't want his wife to know exactly how scared he was—or how little he knew about the threat facing them.

Lisa stood up. "You just remembered something you're not telling me."

James groaned. "Remember that dinner at Sheila's house last spring when we talked about infectious diseases? Daniel Lathrap told the story of how the British gave smallpox-contaminated blankets to American Indians during the French and Indian Wars. And the Indians got sick from those blankets."

"So you can get smallpox from clothing or bedclothes?"

"Yes, it looks that way. But I don't know exactly which stage of the disease is the most contagious..."

His wife interrupted him, hands on her hips. "Did you touch that woman's clothing?"

"I pulled back the edge of her shirt, when I was looking for pustules on her neck and chest."

Lisa's hands twisted into a knot. "Is that enough to catch it?"

"I don't think so. I think the Indians were infected by dried-up pox scabs." What if he were wrong? What if microscopic particles were already in his lungs, poised to spread all over his body?

"There must be a treatment, if you do get it?"

James sighed. There was no point in hiding anything he already knew from Lisa. She was too good at reading him.

"No, and there's no cure. You can prevent smallpox with vaccination, but people don't get vaccinated as children any more."

She gasped. "But I had a shot as a kid—didn't you?"

"Sure, but that was a long time ago. Who knows if I'm still immunized."

He refrained from telling her that once someone was infected, the mortality rate was around thirty percent. He tried to reassure her, and himself. "As soon as the CDC gets here, I'm sure I'll be vaccinated again, just to be on the safe side. We'll be okay. But that poor woman..." He rubbed Lisa's back.

"What now?"

"We wait for the cavalry." James remembered he hadn't meant to touch her and pulled his hand away. "Hmm. Those reported symptoms—fever, vomiting, and rashes—they'll have to test the sick tourists at Alexandria and Petra for smallpox, too."

Lisa gulped. "You think it's connected? But that means bioterrorism..."

He glanced down at her face. Her forehead was dewed with moisture; she was just as scared as he was.

James looked at his watch. "Should we go get some lunch? I'm not particularly hungry after that little adventure, but it will be hours before we get another chance."

"Me either. Normally I'd be starving," she said, trying a weak smile. "But wait a minute, should you even go into the dining room?"

"There's an incubation period, remember. I probably won't be contagious for a few days if I'm going to get it."

"So why can't you stay here with me? The bed is huge." Lisa's tone was acid.

"You know why," he said. "I can't keep my hands off you. And I'm being super protective of you."

"You can say that again," said Lisa, not meeting his eyes.

James locked their room with the big, old-fashioned key, and they walked towards the lobby. As she pushed open the door at the end of the corridor, she turned her head. "James, what if..."

Lisa collided with a thin, dark-haired man her own height.

"Excuse me," he said, and Lisa jerked as if she'd been stung by a bee.

"Greg Manzur! What in the world are you doing here?"

Chapter Ten

Same day, September 10, Luxor

Lisa felt the blood drain away from her face as she confronted a man she had thought she'd never see again.

She could see Greg's shock from the tightened muscles of his jaw and neck. His expression was completely blank.

Then his eyes came alive.

"Lisa! The very last person I expected to see here!" Greg Manzur kissed Lisa on the nearest cheek. His gaze shifted to James. "You must be Henderson. Pleased to meet you at last."

James shifted his weight from one leg to the other. "Sorry? I'm not Tom Henderson. I'm..."

Lisa, her pulse racing, stepped into the awkward breach. "Greg, you wouldn't know this, but I lost my first husband. Tom Henderson was killed in a car accident nearly five years ago. This is my second husband, James Barber." She put one hand reassuringly on James' arm. She glanced at his face and saw the quiver of his lips as he remembered who Manzur was.

Oh, shit. Greg was an old lover from college days. He was also the guy who'd led Lisa into a dangerous race for a lost codex in Israel, when she'd been engaged to Tom. What had begun as a routine trip to secure a museum loan had morphed into a Hitchcock-like chase scene in and out of caves around the Dead Sea. She knew Greg's presence would stir up a hornet's nest of remembered lust, frustration, and fear.

She mentally shook herself as James stuck out a hand towards Greg.

"Manzur. I've heard about you." His tone lacked warmth.

"Good things, I hope," said Greg. "Though I did almost succeed in getting your wife killed, the last time I saw her."

"Yeah. She told me about that little incident. Dodging bullets and hanging off of cliffs."

Greg laughed awkwardly. "A real adventure. But with a happy ending."

"Hollywood stuff," scoffed James, bristling.

Lisa intervened quickly. "Let's not resurrect Israel," she said to Greg. "Are you on this cruise, too? I can't believe you're here just as a tourist."

"Yes and no," he replied smoothly. "I've been in Cairo before, but this is my first trip to Luxor and Aswan."

Lisa snorted. Greg Manzur had fooled her in the past into thinking he was just an itinerant archaeologist, a guy who drifted from excavation to excavation without ever landing a real job as an academic or as a museum curator. Eight years ago, she'd learned he was actually employed by Mossad, Israel's equivalent of the CIA. If he was here as a casual tourist, then she was a world-famous soprano.

The three of them climbed the stairs to the fourth deck and James opened the door to the opulent little lounge adjoining the dining room. They walked past polished brass plant holders and over-stuffed armchairs arranged in inviting groups on a multicolored Turkish carpet.

Lisa tried to remember the different branches of Mossad. When they'd been trying to recover priceless unpublished manuscripts, Greg had been on the illegal antiquities squad. That made perfect sense, given his archaeological background. But jobs changed, and he could have been reassigned. Could he possibly be here investigating bioterrorism? And if he was, then that meant there really was a link between what had happened in Israel and today's illness at Karnak.

The three found seats at one end of a long table and seated themselves warily, Lisa and James on one side, Greg on the other. Oh boy, she thought, a nice little triangle. Wait until I tell Ellen about this—not to mention the fact that James has been exposed to a deadly disease.

Lisa was storing up quite a few things to tell her best friend and colleague Ellen Perkins, but she was too far away for cell phone contact. She'd noticed a public computer in the lobby of their hotel boat, but it was always jammed with fellow tourists checking their email. Maybe there was an Internet café nearby.

The expression on Greg's face was carefully neutral. Lisa, who'd once thought Greg experienced no emotions at all because he was so good at hiding them, now understood that having a poker face was part of his training. It was a survival skill. Lisa wished she could learn that trick—her face betrayed every feeling like a flag. She hoped her skin wasn't as pink as it felt.

Greg looked at James. "So, are you another archaeologist, like Lisa here?" He sipped his iced tea while white-coated waiters passed back and forth taking lunch orders.

Lisa noticed that James ignored his own iced tea in favor of a sealed bottle of fizzy water. Since the ice cubes were probably frozen Nile River water, he was right to be cautious.

"No, I'm a physician, a radiologist. I met Lisa when she was working on Egyptian mummies in Boston."

"Ah! I remember the website about that research: X-rays, CT scans, computer reconstruction—the whole deal, right? Must have been a fun project."

Lisa relaxed. "It was. And that exhibit led to my first real job." She and Greg slipped into shoptalk.

The waiters served them lamb kebabs, rice and plates of salad. Lisa swallowed occasional mouthfuls while telling Greg about the mummy project. It was a safe subject, impersonal enough to distract James, so she waxed lyrical while consuming the excellent meat and the tomato salad garnished with chickpeas and fresh mint. A glance at James showed his face a polite mask, but at least he was eating.

A little voice in her head said, "So now you know what I feel like when you flirt with Betsy!" Lisa squelched the nasty thought and grabbed another piece of pita.

Greg had only taken a few bites of his lamb when he spotted a man in uniform signaling from across the room. He threw down his napkin and stood up. "Excuse me," he said to Lisa and James.

Lisa watched him go. Who was the man who had summoned him? Neither he nor Greg was behaving like an innocent tourist with nothing on his mind except a good lunch.

Then Stanley Rosenberg and Betsy Grover joined them.

"Any news?" asked James.

"No," said Stanley, pouring himself some water. "We haven't heard anything since we left the sick woman's house this morning."

"I'd sure like to know how she got smallpox," said Betsy.

"Keep your voice down!" said James. "We don't know it's smallpox for sure yet, and we don't want to start a panic."

"Betcha it is, and that we'll have media here by tomorrow," said the irrepressible Betsy. "And since smallpox isn't supposed to exist any more, this is probably bioterrorism."

"I happen to agree with you," said Stanley. "But this isn't the time or place to discuss it." As if to emphasize his point, Nicky Reese knocked over a glass of milk at the table just behind him.

"Nicky!" cried his mother. "Try to be more careful."

"Sorry, mommy," said the little boy. He ducked his head but Lisa could see his unrepentant grin.

Greg did not return to the table. Betsy tackled a sticky sweet baklava with gusto, but Lisa pushed her plate away. Stanley and James had just switched their conversation to other infectious diseases when there was a disturbance at the entrance of the boat's dining room.

Lisa looked up. Two white-shirted Egyptian tourist police and a broad man in a gray suit advanced towards them, stopping next to James.

"Dr. James Barber? We have some additional questions for you about the incident at Karnak," said the suited man.

There it was again. Such a little word, "incident." "Potential catastrophe" was more like it.

"Certainly," said James, wiping the honey and pistachios off his beard with his cloth napkin and standing up. "My wife was nearby—she may be able to add something."

The suited man looked at Stanley. "Dr. Stanley Rosenberg and Miss Betsy Grover? You will please come with us also." As James, Lisa, Betsy, and Stanley followed the visitors out of the dining room, several tourists watched and whispered.

"What's going on?"

"Who are those guys?"

Lisa was not surprised that Greg waited for them at the door.

James let Manzur move ahead of him as they followed the officials out of the dining room so he could study this guy from Lisa's past. Medium height, short for a man. Shiny brown hair and a fit-looking body. Manzur didn't look like a threat, but James knew enough about women to know their tastes were unpredictable. Lisa had told him they'd been lovers many years ago, when she was an undergraduate student on a Junior Year Abroad program.

It was what Lisa hadn't told him that worried him. Her recent blush and her fidgeting during lunch, and the gaps in her earlier account had convinced him that the Greg and Lisa affair had been a hot and heavy one, with plenty of residual feelings left over. And they'd shared mortal danger together during her trip to Israel a mere decade ago, always a bonding experience. Was Lisa still attracted to him? Probably. What constituted female sexual attraction, anyhow?

Hell, he couldn't explain exactly why he himself was attracted to one woman and not another—it rarely had much to do with how pretty she was. A nice figure was important, of course. But facial expressions, the way she moved, her voice—oh, especially her voice—all these revealed her personality and her attitude toward the world. His first wife had been delicate-looking, with a sharp mind and wicked sense of humor. James had found that combination devastating.

So James could assume that what Lisa found attractive in Greg was not necessarily the confident set of his shoulders or the tautness of his butt.

Why was Manzur here, anyway? Lisa had told him long ago that Greg was an American by birth, but he'd moved to Israel as an archaeologist and was later recruited by Mossad. Had Mossad sent Greg to Egypt, or was he just a tourist? James tried to remember exactly when the paper had said the Megiddo incident occurred—if the victims had been exposed at the same time as the Farouk family at Karnak, then one could assume a coordinated bioterrorism attack in two countries at the same time. In fact, it looked like the terrorists were already targeting several cities in the Middle East.

The three visiting officials commandeered the cramped but plush administrative office of the hotel boat and quickly shooed out its occupants. The man with the gray suit sat down, motioning to the Americans to sit also, and lit a small, aromatic cigar.

"My name is Detective Hassan," he said. He looked at Stanley Rosenberg. "What is your background, sir?"

"I am a retired physician," said Stanley, his bulky body perched erect on the edge of his chair. "NAMRU—that's the U.S. Naval Medical Research Unit—employed me at the end of the World War II in 'forty-nine. I was one of the physicians working on typhus and smallpox. I stayed on for five years."

"You actually saw the disease, the symptoms, in living people?"

"Several times, in Somalia and Benin."

Hassan turned to James. "You, sir, are an American doctor, that is correct?"

"Yes," said James, crossing his long legs as Lisa squirmed on the sofa beside him.

"I'm a radiologist at a Boston hospital. I also—and this may be of some interest to you—did a six-month stint at the Centers for Disease Control in Atlanta before my residency. And earlier, I was in Doctors Without Borders in Tanzania."

"Ah," said the broad man. "Les Medicins Sans Frontiers, a fine group. Have you ever seen smallpox before?"

"Never. I know what it looks like only from colored pictures in textbooks."

"Ms. Grover," Hassan continued. "You are what kind of nurse?"

"ER," said Betsy. "I mean, I'm trained in Emergency Medicine."

"That may be useful." Hassan steepled his fingers. His brown eyes gleamed as he turned to Lisa. "And you, Madame, are an archaeologist?"

"Yes," said Lisa. "My specialty is Classical and Near Eastern archaeology, but my job is in a museum, not a university. This is supposed to be a holiday trip for us, but I'm doing some research on sphinxes in ancient Near Eastern Art and writing articles for my museum."

The detective glanced at Greg but continued to address his questions to Lisa.

"And you know Dr. Manzur from before, correct?"

"Yes. We intersected twice, many years ago when I was a student on a Junior Year Abroad in Israel and again seven years later, when I was arranging a loan for my Philadelphia museum—not the same museum that I work for now."

James realized Lisa was nervous, because she was talking too much. He saw Manzur nod slightly at the Egyptian detective. He sensed that Hassan already knew all about Gregory Manzur. They must have had a meeting prior to Greg's arrival on the boat at lunchtime.

Greg faced Hassan and the two policemen, his body language indicating a shift in command. "I can vouch for these folks," he said crisply. "Lisa Donahue is an established scholar with connections to archaeologists all over the Middle East. That may help us in dealing with the Supreme Council of Antiquities and the Ministry of Tourism here in Egypt. Barber's and Rosenberg's experience with Doctors Without Borders and NAMRU will be invaluable. And we can always use another nurse."

"Wait a minute," said Stanley. "It's not a confirmed case of smallpox yet."

"It is in Israel," said Greg grimly. "And at Petra."

Oh, God, thought James. It all fit together—first Israel, then Jordan, now Egypt…

Manzur was looking straight at him.

"But I'm not a smallpox specialist," objected James. "I'm a

radiologist. Rosenberg here has the real experience."

"Granted. But you recognized the symptoms before anyone else did, and you're a trained physician. Chances are, when the CDC team arrives, they'll recruit you—and Dr. Rosenberg and nurse Grover—whether you like it or not," said Greg, giving James a very direct stare. "Now, as to why I am here: the government of Israel sent me here right after the Megiddo incident last month because we had a tip that Egypt would be next. A group of terrorists connected to Hezbollah in Lebannon and Hamas in Israel has gotten a hold of smallpox virus from a former Soviet Union stockpile..."

"Wait," said Hassan. He and Greg had a rapid conversation in Arabic. Hassan clearly disagreed with Greg, but Greg overruled him and resumed briefing the four tourists.

"Detective Hassan doesn't want me to put you in the picture, but I have convinced him it is necessary and that you four are to be trusted with sensitive information. However, I must insist that what is said in this room goes no further. Agreed?"

Lisa, James, Betsy, and Stanley all nodded.

"You may discuss the situation with each other, but we don't want to cause a panic among the other tourists, and we want to keep the lid on the inevitable publicity as long as possible."

Nods all around. James reflected that the publicity, when it came, was likely to be intense. The press would have a field day.

Greg continued. "Okay, the goal of this group, whoever they are, is to kill infidels—that is, everyone not of their brand of radical Islam—and cause havoc in the tourist trade. As you know, the tourist trade is a critical part of the Egyptian economy. It is the number one industry in this country. The terrorists first tested the virus in Megiddo, Israel, a few weeks ago, and now we have additional trials like the one here at Karnak. We need to figure out exactly how they disseminated it so we can prevent future attacks. Trouble is, our intelligence from three countries indicates that they are planning something much bigger. We think their goal is to cause a major outbreak here in Egypt."

"So the most likely targets are major tourist traps," said Lisa.

Greg's face was grim. "Yep."

"Which ones?" said Betsy.

One of the tourist police chimed in. "Karnak, the temple complex. The Temple of Luxor. Biban el-Muluk and Biban el-Harim—that's the Valleys of the Kings and Queens. The Temple of Hatshepsut. And maybe even Abu Simbel to the south."

Lisa's pale face shone with moisture. "What about one of these cruise boats?" she asked. "Can you think of a better place to infect a bunch of tourists, all cooped up together in a floating hotel?"

Mr. Hassan bent forward from his ample waist in a little bow. "You are right, Madame. This had occurred to us, also. We must make sure it does not happen or we will have an international incident that will make the Palestinian-Israeli conflict look like child's play."

Chapter Eleven

Saturday morning, September 11, Luxor and Abu Simbel

The early morning air was cool and refreshing, especially after the slightly mildewed smell of the inadequately air-conditioned hotel boat. Lisa walked quickly along the quay, feeling a bit like she was escaping from prison.

She was grateful for the opportunity to be alone (and awake) to reflect on the events of yesterday before being stuffed in another conveyance with her fellow tourists. After the interrogation the day before, she, Betsy, James, and Stanley Rosenberg had found a quiet corner of the deck to talk over their situation.

Stanley mopped his broad forehead and shoved his handkerchief back into his pocket. "Well, I never thought a pleasure tour like this would land me back in a medical emergency."

James nodded. "Yeah. Too much like work."

Lisa said, "How did this happen, anyhow?"

The others looked at her in confusion.

"I mean, how did smallpox virus end up here, in Egypt? If we got rid of the disease worldwide—it was eradicated, wasn't it?—then how come we've smallpox virus lying around where terrorists can get at it?" She directed her question to Stanley, who'd told them the study of bioterrorism was one of his retirement hobbies.

"Good question," said Stanley. "After the official eradication program ended in 1979, there were supposed to be only two places in the world that had a supply of virus."

"The CDC in Atlanta was one of them, right?" said James.

"Right. And the other was the Moscow Institute. But then the U.S. and Britain sent an inspection team into the Soviet Union in the early 1990s and found several virology labs that showed evidence of a recent, very advanced bioweapons program. Russian scientists had developed illegal, weaponized versions of smallpox, both liquid and powdered forms, and had tested special missiles to carry them. But the frozen smallpox virus stores and the biowarheads were never found."

"Didn't some of them end up in Iraq?" asked Betsy, crossing her well-shaped ankles.

Stanley grimaced. "Almost certainly. Although the inspectors didn't find actual weapons of mass destruction, they found facilities capable of producing bioweapons that had been cleaned up in a hurry. And the Russians actually told us that when the Soviet Union busted up in 1991, they had no idea where the remaining stock of smallpox virus went. Some of it probably migrated to North Korea."

Lisa sat stunned, soaking up the horrible details and "what ifs."

They'd gone on to discuss how the World Health Organization had destroyed most of the existing smallpox vaccine because keeping it frozen cost so much. That was before anyone knew about the Soviet program and how it had succeeded in replicating smallpox virus and developing newer, deadlier strains of it.

Lisa now understood why one case of confirmed smallpox here in Egypt constituted a medical emergency. People like her who had been vaccinated as children were no longer protected—it was too long ago. And there were no longer enough doses of vaccine outside the U.S. to treat victims of a major outbreak. Stanley had also described how difficult and time-consuming it was to eradicate the disease. The first time, it had taken years and multiple teams crisscrossing the globe to contain each outbreak and starve the virus into submission. What if they had to do it all over again?

Her companions continued to discuss other forms of bioterrorism.

"Smallpox is one of the two best pathogens for a weapon," said Stanley.

"And the other is pulmonary anthrax," said James.

"Right. Both pathogens are dispersible in air. But to get sick from anthrax, you have to inhale spores, whereas with smallpox, all you need is a few droplets from an infected person's saliva."

"Oh, hell," said James.

"Then there's Ebola," said Betsy darkly. "I saw a case of that once. It's a nasty way to die."

"Ebola has a higher mortality—up to ninety-five percent—but it tends to hit only a few people at a time. Smallpox is easier to spread from person to person. Which is why even one case can cause an epidemic in an unprepared population," said Stanley. "To change the subject, have we figured out new rooms?"

"Yeah," said Betsy. "I already have a single."

"And James can stay with me for tonight," said Stanley. "But I wouldn't be surprised if the CDC pulls the three of us who were exposed to Safaa Farouk off the boat after they arrive and hear the whole story."

This cheerful conversation lingered in Lisa's mind as she walked along the quay.

Soon she arrived at the first kiosk of the Luxor souk.

"Hello, hello, Eee...nglish?"

"What your name?"

"You have children? How many?"

Lisa quickened her pace along the narrow path between stalls. It had seemed such a good idea, getting away from the tour group and her little love triangle with James and Greg. Not to mention the Egyptian authorities. Her mind was spinning as she eyed the merchandise, hunting for something that her nine-year-old, Emma, would love.

A vendor grabbed her arm and threw a cotton dress over her shoulder. "Come! Come!" he said urgently.

She should have dragged her large, bearded husband along as a protector. This bozo was trying to force her into his stall by draping her with unpaid-for goods. Lisa yanked her arm away and tossed the dress back at him. "Not interested!"

The man's assistant caught the dress. He persisted, trying to grab her other arm. "Come! I have good things for you!"

Two more vendors in gray and silver gelabiyahs flanked her. "Postcard, ten for one pound!"

"Nice shirt, thirty-five pounds!"

The accumulated tensions of the past twenty-four hours pushed her over the edge. Time to play her only card. Lisa surged forward with her hands pushing away from her body like a battering ram. "Imshee!" she shouted.

One man fell back. "Why imshee?" she heard him say in a surprised tone of voice. Lisa grinned to herself. People didn't like being told to "get lost!" in their own language.

She broke away from the aggressive vendors and found a stall with colorful embroidered shirts about five hundred feet further along the alley. At last, someone allowed her to handle the merchandise and judge its quality without interference.

Lisa chose a pale azure shirt that looked like it was made for Emma. She took a deep breath and began the bargaining process.

"I'll give you ten Egyptian for this."

"Ah, no! I have to live! I, too, have children to feed! Twenty-five pounds. For you, nice American lady, I do twenty-five pounds."

"That's too much—I have to live, too. Fifteen is all I will pay."

Ten minutes later, her purchase complete, she walked quickly back through the gauntlet of overeager vendors, collecting fake faience tokens and promises of more bargains as she went. She hurried toward the boat, glancing at her watch. Mahmoud wanted them back on board at ten-thirty a.m. sharp so they could be on their bus in time to catch the short morning flight to Abu Simbel.

Lisa reflected that the hard-sell technique seemed to get worse as they moved south along the Nile. What would it be like in Nubia, where they would be in a few hours? She didn't really care. Someone could rob her outright, but no one could force her to part with her cash voluntarily, no matter how many garments he threw over her body or "free" little blue scarabs he pressed into her hand.

The shopping experience, although a bit strenuous, had almost succeeded in distracting Lisa from the fear she'd been sublimating. It was all very well to intellectualize the possibility of a major smallpox epidemic. But when it got personal, Lisa felt the brain cells shut down as the visceral impulses towards self-protection took over. A horrible disease was not one of the souvenirs she planned to collect for her family and friends back home. James had assured her that he hadn't been exposed, but a niggling worry possessed her. What if he were wrong? What if they both got sick, and they never got home to Sam and Emma? After all, as James had explained, hardly anyone knew about smallpox first hand any more. The CDC team supposedly had a plan for controlling smallpox, if it ever occurred again, but they weren't expected until later today.

Then there was the shock of seeing Greg Manzur again. Greg kept showing up in her life like pop-up ads on her computer screen. She'd thought she'd gotten over him once, after her Junior Year Abroad, and then been forced to confront her leftover feelings on that museum trip to Israel only eight years later. Now he was here again, stirring up mud and making James bristle and hover.

As she descended the concrete steps to the Oberoi Hotels boat quay where their cruise ship was moored, Lisa passed the third stray cat she'd seen that morning. This one, a mangy but well-muscled tomcat, ignored her moving feet as he munched on a nameless scrap of garbage.

James, who had been enjoying some fresh air before the temperature soared, met her at the gangplank. "Success?" he asked with a smile, glancing at the small bag under her arm.

"A present for Emma," Lisa said. "At a good price. But I paid in other ways." She told him about the various vendors' ploys as they walked to her room to collect Lisa's daypack and water bottle for the trip to Abu Simbel. "What did Greg say to you after breakfast?"

"Your ex-boyfriend?" Lisa detected a hard edge to his tone. "He told me we should go ahead with our day trip today. He might need some assistance tomorrow around Luxor. But he did say to keep our eyes peeled for any unusual activity. Especially around the ticket booths, souk areas, and tourist cafés."

Lisa scowled as she stuffed her sun shirt into her small pack. She'd wear her tank top inside the bus since its air-conditioning was elderly and intermittent. "This isn't what I'd call a vacation activity."

"Nobody plans on this kind of thing. And if the Pox Gang succeeds in striking a tourist site and some of our group gets infected, none of us will be going home anytime soon."

She stared at him, even as her mind registered how mockingly he'd labeled the unknown terrorists. "You mean we might be quarantined?"

"If someone else comes down with smallpox, everyone who had any face-to-face contact with the victim right before he—or she—got sick will have to be isolated."

Lisa slung the pack over her shoulder and picked up her oversized metal room key. "So we're lucky no one in our group is sick—yet. Or that the rest of the tour isn't canceled. I can just imagine the complaints Mahmoud would face if he announced that all these tourists were not going to get their money's worth."

"Yeah, it would be hell on him." James opened the door to the lobby for her. "Get the most out of today. You never know, we might be stuck in Luxor for a while."

He lowered his camera to watch his wife a couple of hundred yards away. Lisa stood perfectly still, her gaze focused on the gargantuan statues facing them. Her head tilted slightly so she could see out from under the broad brim of her sunhat. Yup, she was mesmerized. Her body was still, but he sensed it was like a tightly coiled spring. Was it the excitement of finally seeing monuments she'd only read about before, or the sudden encounter with a man from her past?

James hunched his shoulders up and down deliberately, releasing tensions. He focused his mind away from his personal fears to the incredible sight in front of him.

They were now about forty kilometers north of the Sudanese border, in the region known in ancient times as Nubia. James and the other tourists stood on a broad terrace, flanked on one side by a sprawling wall of sand-colored stone and facing the huge, carved niche of the Great Temple of Abu Simbel on the West Bank of the Nile. The four cloned pharaohs (Ramses, Ramses, Ramses, and Ramses) inside the niche were, without a doubt, the biggest statues he'd ever seen. And the queen Nerfertari—or was it Nefertiti?—standing between the pharaoh's feet was so much smaller.

"He must have had a really big ego," said Stanley Rosenberg, who appeared next to James. James guessed that he was deliberately avoiding the subject of bioterrorism and pretending, for the moment, that they were just ordinary tourists.

"Yeah. Enormous, like all his works," answered James. He remembered another statue, a pink-and-black granite one, lying on its side in a museum in Memphis. Also Ramses II, with amazing hieroglyphs on his wrist that James had photographed in living color and digital detail.

The retired doctor motioned to James' camera. "That an Olympus? What model is it?"

"A C-5050," replied James.

"With a zoom lens?"

"Yeah."

"How do you like it?"

"It's great—a little heavy for foreign travel, though. I got Lisa a point-and-shoot model that's tiny. But I can take close-ups of hieroglyphs and stone texture that she can't with hers."

"Can I see it?"

"Sure," said James, handing it over.

"I'll have to look into Olympus before I buy my next one," said Stanley, after he studied the camera body and its various buttons. He changed the subject. "Isn't it amazing that this whole site was buried, excavated, and then moved, block by block?"

"Yeah. It's all connected with the building of the High Dam in the 1960s. They moved the Philae temple too, from one island to another." Lisa had told James that the Oriental Institute in Chicago had a whole basement full of artifacts rescued from the rising waters. She often said one could make a career out of "museum excavation" because most major museums had huge amounts of un-researched, unpublished material in their storerooms.

Stanley Rosenberg broke into his thoughts again. "I heard the

Centers for Disease Control team is on its way."

"Then all hell will break loose," said James. He gulped and ventured an awkward question. "Hey, Stanley, remind me, what's the incubation period for smallpox?"

Rosenberg's eyebrows shot up. Clearly, the other doctor guessed how his mind was working. "Seven to fourteen days, depends upon the individual."

That fit with what he remembered. James felt his throat go dry. "And the most infectious stage?"

"Before the victim has any pustules, but has a fever, headache, nausea, etcetera. That's when little droplets from a cough could infect you."

So Safaa Farouk was—most likely—at her most contagious only a few days before James had seen her.

"But some sources say the virus can be caught anytime in the first three weeks—from initial symptoms until when the scabs fall off. That's less likely, though," said Stanley.

Great, thought James. Nobody really knows what we're dealing with here!

Stanley moved into position to take his own photos while James, beads of sweat rolling down his back under his shirt, replayed how he had leaned over Safaa Farouk, breathed the same air.

Suddenly he noticed that Lisa had stiffened like a hunting beagle catching a scent.

"James!" she called, whipping her blond braid around her head as she turned and walked towards him.

"What is it?"

"I'm not sure. But our bus driver seems awfully thick with one of the local tourist police. They've been standing together for a good ten minutes now, and I thought I saw the driver pass him something."

James frowned. "Are you sure it wasn't just a cigarette? All these guys smoke like chimneys."

"No, I'm not sure of anything. But Greg wanted us to report everything, right?"

"That's what he said." James was already a little tired of Manzur telling them what to do. He watched the bus driver leave the tourist police and saunter over to the outdoor café. Colorful drapes with patterns made out of Arabic script adorned the sides of the tent-like structure, allowing what little breeze there was to pass through and cool the over-heated streams of American, Japanese, and German tourists. A vendor did a brisk trade in Coca-Cola and bottled water at the entrance, while inside another man sold gooey-looking pastries and chocolate bars.

The MISR security guard stood near the drinks stand. Zawi Farouk was tall, with Espresso-colored skin and a striking nose that was slightly broadened at its base. A mix of northern Egyptian and southern Nubian ancestors, perhaps? His suit was well cut but fell loosely over his muscular body. James figured that most of his fellow tourists didn't know about the gun Farouk carried—James had seen it only because he and Farouk had been in a men's room at the same time, and the other man had briefly shed his coat.

Farouk caught James watching him and nodded without smiling.

James inclined his own head and pulled out his handkerchief to mop his sweaty brow. At least the tour group had one guard who not only was armed but who looked as if he would be competent in a fight.

He resumed his slow stroll around the site while his thoughts raced. There were so many questions he had about smallpox. Naturally he had no textbooks with him, but that wouldn't matter if he had access to the Internet.

James wished he knew exactly how smallpox virus was weaponized. He had never seen a report on how it could be done. Web access probably wouldn't give him that recipe, if it existed. Multiple countries would try to keep that information top secret. He knew the virus was amplified—grown—in a kind of biological soup in a bioreactor, which could be both small and portable. Stanley had told them the Russians had figured out two ways to use it, as a dry powder and as a liquid. They had found a way to keep it frozen in special warheads so that missiles could carry it without killing the live virus.

All those experiments with different smallpox strains. What if the scientists had developed a mutated, super pox that no vaccine could cure? Or what if they'd made a form of smallpox that was as contagious as the common cold? He shuddered and turned his thoughts to the equally pleasant topic of how the terrorists might engineer a major outbreak.

They'd use the simplest means possible. They'd spread the virus, most likely as a fine powder that would carry particles into human lungs. The terrorists could pass vials of the stuff from one individual to another. Each man would be instructed to open or break the vial at a time when the maximum number of tourists would be infected—preferably in a closed air space like a bus, a train, or a sealed building.

James looked around him, pretending he was a terrorist scoping out sites. What about here? That café was right next to the bathrooms and on the edge of the souk, both tourist traps. And the market was cleverly positioned between the archaeological site and the bus parking lot, so all the tourists had to run the gauntlet of numerous vendors bent on relieving them of their foreign cash.

James was no policeman, but he was a prolific and eager photographer. Then and there he vowed to take as many pictures of native security guards, ticket-takers, tourist police, and bathroom attendants, as his camera storage card would hold.

Zawi Farouk was bored. This job was really the limit. These long periods of inactivity, just hanging around while the tourists gawked at the sights and revealed their poor Western manners and their aging flesh to the natives.

He lit another cigarette and leaned against the wall of the temple enclosure wall, letting his thoughts wander to more interesting topics like how much money his moonlighting in terrorism would make him. Why, he could build an addition to his house. Maybe he could talk Achmed into moving in with Zara after they were married. No, Achmed was in his forties, and like Zawi, had a perfectly good house of his own. Besides, he already had two wives of the allowed four wives permitted under Muslim law. And if there was bickering among the wives, as there inevitably was when a pretty newcomer was added, Zawi wanted no part of it.

He shoved aside the memory of his daughter's huge, tear-sodden eyes, pleading with him to let her marry that good-for-nothing pipsqueak, Farzad. He refused to feel guilty. After all, he was the father, the head of his household, and Zara would just have to learn to adapt to her changed circumstances.

Zawi ground out the cigarette under his heel and strolled over to the bus, waiting on the edge of the parking lot. By now Ali, the Nubian bus driver, should have collected his package from the local agent.

"How are you, my friend?" asked Ali in Arabic.

"Alhamdulallah (fine). You have something for me?"

"Oh, yes, here it is." Ali slipped him a little box, wrapped in brown paper, and Zawi slid it inside his suit jacket. The jacket was custom-tailored not only for the automatic pistol and rounds of ammunition he needed for his MISR Travel job as a security guard but had plenty of pockets for less legitimate cargo.

Ali told him an uncomplimentary joke about American women and Zawi's lips split in a cynical smile. Then, as his gaze moved over to where the American couple, Dr. Barber and his wife were standing, he noticed that both husband and wife were looking at him.

This was not good. They weren't behaving like disinterested tourists at all. He had already noticed Ms. Donahue wandering into alcoves and storage areas where no other tourist bothered to look, and her husband had the unfortunate habit of photographing everything. When Zawi had asked him yesterday why he took pictures of the ground, Dr. Barber had told him he liked texture—in the dirt, in the rock, in any scrubby vegetation by the side of the road—all these were photographic fodder for a doctor trained to recognize patterns.

Zawi moved away from the bus and crossed to the café and restrooms, where most of the MISR Travel group was already congregating. Out of the corner of his eye, he continued to observe James Barber. Blast the man, now he was taking photos of some of Zawi's own employees, who also worked in the tourist trade at Abu Simbel!

He would put a stop to the picture taking, and quickly. Zawi strode rapidly to the Coca-cola kiosk, where Jawad worked. Jawad was fleet of foot and very, very good at petty theft.

"Jawed!"

"Yes, Boss."

"I have something for you to do. First, find a replacement at your kiosk. The task I have in mind will take only a few minutes."

Dr. Barber was about to lose his precious camera.

Chapter Twelve

That afternoon, Abu Simbel and Luxor

"Smallpox was—and is—one of the worst diseases in the history of man."
(from the diary of Stanley Rosenberg, M.D.)

Soon, the tour group was on its way home from the day trip to Abu Simbel. They were about to board the bus that would take them to the airfield, and then a short flight would return them to Luxor.

Many of the group, including James, were still clicking away with their cameras and making last-minute purchases in the souk. A hot and sweaty Lisa joined the first wave of tourists clambering onto the bus. She reflected that it had been a successful day packed with interest and with no unusual incidents—almost like a vacation after all.

The aging bimbos from Iowa were in full spate, chattering and giggling over their latest purchases. They had forgotten Bertha, marooned back in Cairo, and were not in the least impressed with the most spectacular temple in southern Egypt. Or rather Nubia. Lisa took a restoring swig from her water bottle, leaned against her seat, and closed her eyes.

Abu Simbel or Karnak? She couldn't decide which was the best site. Abu Simbel won on size of the sculpture—those gargantuan statues of Ramses II—but Karnak had everything: narrow passageways, obelisks, sudden grandiose views, giant scarabs, and a lake. On the other hand, Abu Simbel had such a story. The entire complex was moved between 1960 and 1980, and most tourists didn't even notice how neatly the cut blocks were fitted together until a guide pointed them out.

Lisa opened her eyes and looked out her window. James was approaching the bus, walking at his usual relaxed pace with his camera bag slung over his shoulder.

She saw a dark young man wearing tan slacks and a white shirt sidle up to him. Suddenly he snatched the camera bag off James' shoulder and nipped around the front of the bus.

"Hey!" yelled James. He took off after the thief.

Lisa jumped up and crossed the aisle, leaning over Patsy to see out of the window.

"You've got your own window, for heaven's sake!" said Patsy.

"Sorry. My husband just had his camera stolen, and I was trying to see which way the thief went," said Lisa.

"Oh!" cried Patsy, and immediately everyone nearby strained to see.

Lisa sat down again as she waited for him. The thief had undoubtedly hidden himself near one of the other tourist buses—easy to do since there were about fifty of them parked near the entrance to the site.

James reappeared, looking dusty and disgusted. When he climbed on board the bus, he was immediately assaulted with questions.

"Did you catch him?"

"Can the Tourist Police do anything?"

"No, and no," said James. "There are too many tourists, too many buses, and I couldn't even find the Tourist Police." He plumped down in the seat beside Lisa.

"You had a lot of pictures on that camera," said Lisa softly.

"Yeah," He looked at her. "I know you have duplicates of the major sites, but I had some really great detail photos."

She squeezed his arm as the bus started to move. "We'll get you another camera."

James slumped a little lower in his seat and scowled. "You know, sometimes I really hate traveling."

"I agree. Getting robbed and getting sick takes a lot of the fun out of it," said Lisa. "But think of the stuff we've seen! Even without all the pictures, the memories are priceless."

He managed a smile for her. "I suppose you had a good time looking at all the archaeology?"

"And how," she sighed. "This is what I came to Egypt for. Wouldn't it be nice if we had a few more ordinary, touristy days with nothing more exciting than crawling in and out of tombs?"

"I doubt it's crawling any more," said James. "I bet you go in and out by graded ramps and wooden walkways, like the ones at the Pyramids. But I can see you'd prefer the more romantic method of tomb visiting—trying to walk bent over like we did at that Pharaoh's tomb at Giza and finding your way carrying a flashlight with dying batteries."

Lisa grinned. "Yep. Or at least that was what I wanted ten years ago. Now, as I approach sober and creaky middle age, I prefer walking upright with electric lights strung out in front of me."

The bus pulled up to the steps leading down to the Oberoi hotel boat Shehrayar. Through the bus window, Lisa saw Greg Manzur hovering near the ramp and smoking a cigarette. He looked straight at her but did not smile.

"Uh-oh. I think Greg has more news for us."

Lisa was right. Greg had news, and none of it was good. He herded them upstairs to the boat's passenger deck, correctly assuming that none of the other tourists would follow them. Lucky them, they were headed for hot showers and cool pillows. Lisa, feeling grubby and resentful of Greg's abrupt reminder of things she'd rather forget, flung her knapsack down on one lounge chair and flopped down on another. She scarcely noticed the softening evening light and the silent felucca sailing past as she glowered at her ex-lover.

"Okay, what's so all-fired important that you can't even let us go to our room and freshen up first?"

Greg made no apology. Instead, he said, "The Egyptian woman—Safaa Farouk, the one you examined, James—is dead. And the test came back positive. We have a confirmed case of smallpox on our hands here in Egypt." He paused. "And Bertha Miller is getting worse. The hospital in Cairo called to say her fever has dropped, but she now has a pustular rash on her face and torso."

"So that's probably another case of smallpox—from our own group. But confirmation will take a few days—they'll have to send her swabs and blood samples to a lab that can handle biohazards." James clenched his hands between his knees and leaned forward in his chair. "And the CDC? Are they here?"

"In spades. The Centers for Disease Control had sent a team of fifty docs and nurses and administrators, plus all the stuff they need to set up a field clinic and vaccinate people. Their representative will be here on the boat in a quarter of an hour. He wants to talk with you."

Lisa felt the sweat of fear trickle down her torso. "I'm sorry, Greg. I didn't mean to snap at you. What will happen after we talk to the CDC?"

A corner of Greg's mouth curved up in a half-smile. "I don't know, exactly. I predict, however, that the next two days will be a lot more chaotic than it was in Israel with the first outbreak. There, the Israeli Army took over and things moved quickly. Here, I think there will be considerable backing and filling and recriminations before they figure out who's really in charge. And there's something else you should know."

Lisa and James waited.

"Two other members of the Farouk family are sick."

"Same symptoms?" asked James.

"Identical. A teenaged boy came down with a high fever that went away after a couple of days. Then he developed spots. Now his other aunt is sick."

Lisa said, "Farouk? That's the name of our security guard, isn't it? Is it the same family?"

Greg's dark eyebrows shot up. "Yep. One of the first things we checked."

Chapter Thirteen

The same afternoon, Luxor

"Verily, the city is destroyed by death from pestilence and disease."
(Amarna Letters, No. 244)

Abdul Farouk lay on his sweat-soaked sheets, drifting in and out of sleep punctuated with horrible dreams. The dreams faded into shouting and wailing, and dimly he realized that he was awake. A fly buzzed around his head in the windowless room.

"She's dead? How could she be dead?" cried a voice in Arabic. It was his mother, Mina. "And my son, my baby, he is sick too!"

"I cannot understand it," muttered Uncle Ali.

Abdul heard footsteps approaching the kitchen and then a clattering sound.

"What! This is not the way I left it! Who has been in my blue jar where I keep my special tobacco?" thundered his uncle's voice.

"No one, no one," soothed Mina, who was usually the one who could calm Uncle Ali when he blew up in a white rage.

Abdul heard clicking as his uncle rushed in through the bead curtain that separated the sleeping room from the kitchen. He opened his gritty eyes and saw the purplish face of his angry relative.

"You!" cried Uncle Ali. "I know you like sniffing my tobacco! You stuck your filthy hands in there, didn't you? You took the package out! Didn't you?"

Abdul tried to moisten his cracked lips. "I never…I did not think…" he gasped, as his mother Mina rushed into the crowded space and plucked at the long sleeve of Uncle Ali's gelabiyah.

"Think! Of course you didn't think! Now your aunt is dead, you meddlesome son of a jackal!" Uncle Ali bent over Abdul and grabbed his shoulders with strong fingers that felt like steel pincers.

Mina shrieked, "Let him alone! My poor boy is sick, can't you see? Maybe he will die too from this horrible sickness!" Then her eyes opened wide with horror. "He has the spots! Just like my sister Safaa…we will all die!" She covered her face with her headscarf and groaned.

Uncle Ali's face was no longer purple. It was a curious shade of white, tinged faintly with blue, as if he felt suddenly ill himself.

"It cannot be," he whispered, wiping his hands on his robe. "They didn't tell me it would be like this! My own family."

The bead curtain swayed and clicked as Uncle Ali slunk from the room.

Abdul turned his aching head toward the mud-plaster wall and closed his eyes.

He woke up again to the sound of raised voices.

"I cannot, Papa! I cannot marry that man! He is old and fat and ugly. Besides, I love another."

"Phaaah! That Farzad Hammadi, he is nothing, worth nothing." It was his papa, Zawi Farouk, talking with his sister, Zara. "What is his father? A seller of T-shirts! He has no education; he will never amount to anything. Jebal Allah Achmed is established in his rug factory, a substantial businessman."

Abdul heard his sister scrape back her chair. "I won't, I tell you! I'd sooner die first!"

"You will do as I say! And don't be ridiculous, my daughter. Achmed will be good to you. Why, he has told me he will add to your dowry and he will pay for the wedding."

Zara shrieked her outrage. "He will put more money in your pocket, you mean! What kind of father are you to sell his youngest daughter like a camel, like a donkey?"

Abdul heard a slap and then a gasp.

"You will not speak to me that way! I am your father, and I decide! The wedding will be in one week."

He tried to sit up, to go to his sister. He wanted to tell his father he must not do this terrible thing. But Abdul's muscles would not obey him. His arms were weak, and he was so dizzy. The sickness overcame him, and he flopped back down on the bed.

Now all he could hear was Zara's weeping. His father had left the room, leaving his sister to master her rage and grief alone.

Abdul croaked out her name. "Zara! Come in, come in to me!"

The beads clicked and she was there, her face swollen and reddened from crying. Even then, she was beautiful.

"You heard, my brother? He will give me to that misshapen monkey, that miserable old fart!"

"I will not let him. I will talk to him, just as soon as I am better."

Zara laid a slim brown hand on his arm and stroked it. "It is useless. We both know Father. Once he has declared his mind, nothing will change it. Oh, I wish I were dead!"

"Do not say such things, dear sister. It is not lucky. Go to your Farzad; maybe he will help you. Maybe he can move the shop to another town, away from our family. You could learn accounting. Then I could come visit you, maybe help in the shop, until I get into medical college."

She leaned over and embraced him with no thought for his sickness or her own safety. "I am very lucky, Abdul, to have you as my brother."

Abdul clutched her like a lifeline. As he released her, he heard voices at the front door.

"What? He cannot go. I am his mother, I will nurse him!"

"I am sorry, madam. Your son must be moved to the field hospital because he is contagious. We cannot let him stay here. He will make all of you sick, too."

After more discussion and protests from the Farouks, two tan-uniformed men, one American and one Egyptian, pushed through the bead curtain with a stretcher. Despite the combined protests of Zara, Mina, and Zawi, the men took Abdul out to the ambulance.

As the medics began to load him and his stretcher onto their truck, Abdul looked back at his family standing in the doorway.

The American, who spoke Arabic, turned back to the Farouks. "Someone will come to interview you all again, to ask questions about the sickness. Now, did all of you accept the vaccine when Safaa Farouk was diagnosed?"

"My wife and daughter took it. Abdul took it also, but he was already feeling sick. I, and most of the men of my house refused your Western medicine," growled Zawi.

"Your sister-in-law is dead. Your son may die," said the medic. "I think you should reconsider. The vaccine is very effective, and it is for your protection."

"Pah!" said Zawi. "I do not change my mind because an American says so."

The medic sounded impatient. "Then those of you who are not vaccinated must stay here, in the house. You must tell us if anyone else develops a fever or a rash. The guard we will send is for your safety and for the safety of all your friends and relatives. He will keep other people from visiting, people who do not yet have the vaccination. Only those of you who have been vaccinated can go into the market to buy your food and other items. Do you understand?"

Zawi said nothing while his wife nodded.

Abdul closed his eyes. Through the pounding in his head came the thought that Zawi had certainly understood, but was unlikely to obey.

The Luxor souk was alive with color and noise. Zara walked quickly, her tear-streaked face partially hidden by her headscarf. It had been easy to leave the house since there was no guard yet; besides, she had accepted the vaccine despite her father's tirade against all things Western. She would find a way to visit her favorite brother.

Her mind seethed with visions of escape using Farzad's motorbike. Even as she envisioned taking off at three in the morning, she knew it would never work. They must have a real plan—one with a destination well out of her father's reach. And they must have enough money so that they would be beholden to no one until both of them had jobs. The few coins she was able to scrounge from shopping for her mother were totally inadequate for an elopement.

Gulping back a sob, Zara slipped into the alley that led to the back of the Hammadi alabaster shop. Farzad usually worked in the back, packing alabaster statuettes for shipping. Surely he would have some ideas; they had already begun to talk about where to go if Zara's father proved impossible.

She spied his hawk-like profile through the bead curtain.

"Farzad!"

"My darling! I didn't know you were coming so early."

She gripped both his arms and looked imploringly into his face. "I had to come. Listen, my father insists that I marry that old baboon, his business partner. It will be soon, a few days or a week. As soon as he can make the arrangements."

Farzad's mouth set in a hard line. "Never! Not while I have breath in my body! Okay, we must elope, then. We must find somewhere to stay, though. Maybe Esna or Edfu."

"But is that far enough away?" asked Zara, shivering. "My papa, he will be very angry. I think he will follow us."

"Not if he has no idea where you are! And my family, they approve of you." His black eyes shone with earnestness. "They will not give us away."

Zara was not so sure. Family tradition was so strong. People hardly ever allowed young people to do as they liked; they were always caught in a web of shoulds and musts. She had a feeling Farzad's father would think he owed it to Zawi Farouk to retrieve his daughter...but maybe not. They were not friends, but competitors. And Zawi's reputation was that of a violent man. Perhaps the Hammadis would keep their mouths shut for Farzad's safety if not her own.

She smiled tremulously and released Farzad's arms. "Will your motorbike last long enough to get there?"

Farzad bridled. "You doubt my machine? My pride, my joy? Of course she will make it!"

Zara really smiled this time. Why did all her male relatives refer to their bikes and cars, no matter what they looked like, as female? It must be the sense of possession, of control. After all, everything else female was some man's property.

"When shall we go?" She used a corner of her headscarf to scrub away the tearstains.

"Has your father set a date?"

"No, he must finalize the agreement with horrible Achmed," Zara's tone was bitter. "And work out exactly how much money Achmed will pay to take me as his bride. I feel like a camel or a donkey being bartered away."

Farzad stroked her smooth cheek. "Do not worry, it will never happen. I will find a way." He reached for her but then checked his movement as a woman stepped into the alcove. "Look, your mama is here!"

Zara pulled away from Farzad. "Mama! What are you doing here?"

"I have come to help you, my daughter." Mina's eyes glowed with affection and determination. "I have come to talk with Farzad's father and to propose how to get you both away before my husband locks you up in a loveless marriage."

Zara stared at her mother in wonder. "You, Mama? I know you support our love, but how can you help us?"

Mina sank into the only chair. "I have a second cousin in Aswan," she said. "I think he could be persuaded to house you until you find your own place to live. Also, he has connections in the local souk. He can help you locate a shop to sell your alabaster and other goods."

Farzad smiled broadly. "But how wonderful! Why should this cousin do this for me, a young man he has never met?"

Mina sat straighter. "Because once, a long time ago, he wanted to marry me. He will help my daughter and the man she wishes to marry, for my sake."

"Mama!" cried Zara. "Did you love him?"

"Yes. But my father would not listen, and so I was sent away to marry into the Farouk family. And so Moustafa found another woman."

Her lips twisted with sadness.

"You see!" cried Farzad, pulling Zara into an embrace. "You are going away with me and we will be very, very happy."

Mina Farouk watched them and smiled. She hauled herself out of the chair and headed for the main shop. "Now I will talk with your father, Farzad. Zara, don't stay too long. It is best you do all your chores on time so your father can find no fault with you."

"Yes, Mama."

Chapter Fourteen

Sunday morning, September 12, Luxor

The officer in the white uniform argued with his Egyptian counterpart. "The Centers for the Disease Control in the U.S. has sent us. We have a protocol we must follow."

"But this area is not under your jurisdiction," replied the thin man from the Egyptian Ministry of Health.

"It doesn't matter what the normal protocol is—the CDC is in charge now," said the officer, now red in the face. "We have brought our Emergency Response Team, a mobile clinic, the works." He stepped back as a tan-uniformed member of the Supreme Council for Antiquities arrived at his elbow.

"Please, I am Dr. Zahi Hawass' assistant. He insists we will all have a meeting in the tent over there." The assistant smiled, a thin, stretched grimace that did not reach his cold eyes. "You too, Drs. Barber and Rosenberg. This way."

James, who had been listening to the exchange, joined Stanley and a jostling throng of men at the entrance of the tent. The tent had been hastily erected the previous night near the tourist entrance to Karnak. Beyond the billowing white canvas rose obelisks and thick pillars of stone, an incongruous setting for this meeting of worried men from Egypt, the U.S., and Israel.

They were in for a difficult time, James thought. With the number of different agencies representing three nationalities, the real challenge would be to agree on any protocol at all, let alone one that would actually control the situation. The damage had already been done. A confirmed victim of smallpox had died, and no isolation of family members had yet been implemented. If somebody didn't move very fast, any efforts made by an international alliance of agencies would be ineffective. The disease would certainly spread.

And James had been exposed—possibly twice. If Bertha's tests came back positive, then he, Stanley, and Betsy were doubly vulnerable. He squirmed as he recalled what the CDC representative had just told him. All people who had been within six feet of a smallpox patient showing symptoms were considered "face-to-face contacts." He could imagine Lisa's horrified reaction to this news and he hoped he wouldn't have to tell her.

He and Stanley took their places, sitting on folding chairs provided by the Egyptian branch of the World Health Organization. As Betsy Grover joined them, James spotted Greg Manzur at the front, talking to two men in uniform.

An imposing Egyptian clapped his hands and drew the attention of every man present.

"I am Dr. Zahi Hawass, Secretary General of the Supreme Council for Antiquities," said the Egyptian. A thin young man translated his English statement into Arabic. "I represent the government of Egypt in this matter."

An appreciative murmur hummed around James. Then the man sitting in front of him hissed at his neighbor, "What is Hawass doing here? This is a health matter, nothing to do with antiquities."

"It involves dangers to tourists," said the neighbor. "And anything involving tourists is big money to the Supreme Council and to all Egypt."

"We have here representatives of the Ministry of Tourism, the Centers for Disease Control from Atlanta in the U.S., the World Health Organization's office in Cairo, and the NAMRU agency that has a branch in Cairo," continued Hawass. "We also have an agent from Mossad in Israel, who knows the circumstances of the first outbreak of smallpox, and two American doctors with expertise in tropical diseases." Hawass nodded towards James and Stanley.

James immediately felt uncomfortable. Just because he'd seen service in Africa years earlier and had helped Stanley with the initial diagnosis was no reason to imply he was an expert in tropical medicine!

Hawass said, "I will now turn this meeting over to Dr. Thomas Mallory of the CDC. He will describe protocols developed by that agency for dealing with this type of situation."

James sat up straighter. This was a good sign, if they were going to let the CDC take over. But it didn't prove to be that simple.

"One moment, sir," said a portly, olive-skinned man in a crumpled linen suit. "The World Health Organization's Egyptian branch should speak first, as we already have some procedures in place."

Another Egyptian said, "The Ministry of Health should really be in charge here."

Several people stood up and started shouting.

"The victim's relatives, where are they?"

"Has anyone started the vaccinations yet?"

"The CDC has procedures we must do..."

"Where is the doctor who treated the first victim?"

Chaos ensued for several minutes until Mallory, the CDC representative, finally managed to get everyone to sit down. "I need your attention up here, folks. Now I have a piece of news most of you have not heard yet: a German tourist on the cruise ship Sun Boat III presented with smallpox-like symptoms on the same day the Egyptian woman was found. His tests came back positive early this morning, so he is the second confirmed smallpox case, and there are two more in the Farouk family. Finally, there is Bertha Miller in Cairo, but her case is not confirmed yet."

"Are you quarantining that boat?"

"What about all the other tourists in Luxor?"

Mallory held up his hand again. "We'll get to that. First, I'd like Drs. Stanley Rosenberg and James Barber to come up front and describe to us the first case, the Egyptian woman Safaa Farouk, and what has been done so far. Dr. Barber, let's have you first since you made the discovery of the disease."

James, feeling the sweat gather on his upper back and roll down his skin beneath his shirt, repeated his story. "Safaa Farouk lives in a large house just off the Avenue of the Sphinxes near the archaeological site of Karnak. When I visited her, she was at the pustular stage of the disease." He spoke for several minutes more, finishing with, "I have no idea how many people actually live in that house, or how many were in recent contact with her." He sat down in a rickety chair and his colleague rose.

Stanley Rosenberg pushed his thick glasses higher on his nose and tucked in his shirt. "Dr. Barber called me in as soon as he noticed Mrs. Farouk's symptoms. I agreed that it was probably smallpox. Now, the first thing we must do is find the rest of those relatives," he said. "And interview the other tourists on the German cruise ship. I saw smallpox on my tour of duty with the U.S. Navy NAMRU unit back in forty-nine, and this is how we handled it..."

After Stanley finished his account, another doctor was asked to report on the sick German, a Viktor Frankel. Then the discussion moved on to how to contain the outbreaks.

"Exactly how contagious is smallpox?" asked a young nurse wearing a CDC identification tag.

Mallory answered. "It depends on which stage of the disease it is. Initially, the victim has no symptoms and is not contagious. Then, he develops a high fever, lots of aches and pains, and is essentially prostrated. That's when the virus begins to build up in the blood and bodily secretions, especially the saliva." He went on to describe the most contagious state—when the rash emerged, beginning in the mouth and then spreading to arms and legs—and how the rash changed to the characteristic pustules that James had seen on Safaa's neck and chest. James was relieved to hear—again—that the pustular stage was not as contagious as the rash immediately preceding it. And Bertha Miller's fever was just beginning when he'd examined her at Giza. He began to breathe easier. With the vaccine that he'd just received coursing through his body, James might escape serious problems.

"In really bad cases, the pustules merge into confluent sheets all over the body, and the skin tears away from the under layers. The face can turn into a bubbly mass full of fluid. If the victim survives, he or she is often left blind," continued Mallory.

James shifted on his hard chair.

"How is smallpox weaponized?" asked the same young woman from the CDC.

"Well, we think the initial amplification of the virus is done in small bioreactors like the ones the Iraqis had in the 1990s."

"How big are the bioreactors?" called out someone at the back of the tent.

Dr. Mallory smiled, suddenly human. "The inspectors in Iraq saw small ones, about the size of R2-D2, that little robot in Star Wars." He continued, "Next, the virus is dried and converted to small particles that can adhere to clothing and other fabrics. We don't understand this process very well."

"They must have used some kind of binding agent to make the particles sticky," said another doctor.

"We may never know exactly how they managed it," said Dr. Mallory. "So, let's get back to how to treat the disease now that we have two confirmed cases."

Intense discussion followed, while the various officials tried to sort out which government's protocol to follow and who should be in charge.

A tall, slightly balding Egyptian stood up and commanded their attention. "I am representing the Ministry of Tourism," he said. "We don't want the international press to find out about this situation. It will be very bad for tourism in Egypt, very bad."

"It's a bit late for damage control," said one American doctor. "You can't keep quiet about something like this now that several agencies and three governments know about it."

"Nonetheless, we must try," insisted the tall man. "We must limit discussion outside this room."

James reflected that the word "room" hardly applied to an outdoor tent. All the physicians and government officials had IDs, and there were armed guards at both entrances, but what about the dozen or so service personnel wandering in and out serving coffee and pastries? Someone would talk.

Finally Thomas Mallory stepped in again. "I recognize that we have several different agencies here and three different governments. While someone more appropriate may appear later, I am volunteering myself to take over the medical side of things while Dr. Hawass here deals with the local tourism folks and the police. Time is critical—we must jump on this disease as quickly as possible, complete the interviews with everyone who has had any contact with Mrs. Farouk or Mr. Frankel, and begin vaccinations. Sun Boat III must be quarantined. We must also interview all the tourists on the Shehrayar, the tour boat moored next to the Sun Boat III to find out how much those two groups of tourists have mixed. They've had exactly the same itinerary over the past week. Agreed?"

Hawass looked as if he'd rather be the one in charge of the whole operation, but he bowed slightly in acknowledgment of Mallory's emergency medical training.

The meeting broke up into small groups, and Dr. Mallory pulled James, Stanley, and Betsy aside. "I'd also like to ask you three to join my team and move into the hotel with the rest of us. All three of you should be vaccinated immediately, anyhow, since you had close contact with the patient. Then you will be as well protected as anyone else on the medical team and can help us conduct interviews with the locals. Okay?"

Betsy and Stanley nodded their agreement.

James frowned. "I've just had the vaccine. What about my wife? She's still on that tour boat."

"We'll be grounding the tour at Luxor until we conduct the rest of the interviews. With ninety passengers, that should take some time. It's helpful that no one else on your boat has been directly exposed besides you three," said Mallory. "And Dr. Rosenberg had the sense to suggest that you and your wife sleep separately last night—but of course, you are not contagious yet, even if you were exposed to Safaa Farouk." He stepped to one side and spoke briefly with one of his officers. "I think we can let the tourists visit other sites in the immediate area, but the boat cannot leave the quay."

James was relieved. If he had to be quarantined, at least he would be busy. Lisa would be upset at his kidnapping by the medical establishment, but she'd understand that he had no choice in the matter. She'd still be able to see the sites they'd traveled so far to visit, and Lisa could continue her sphinx and travel articles. His reflections were abruptly terminated by Mallory's next statement.

"Of course, if anyone on the Shehrayar cruise ship develops a fever or other suspicious symptoms, that's especially serious because it's a closed air space. We'll have to quarantine and vaccinate everyone on board. Tourists, crew—the lot. No one will be allowed to go on and off that boat except medical personnel in protective gear."

The man with the olive robe and white kefiyeh wrapped around his head served coffee to the meeting attendees. This gathering was all Western suits, hardly any Arabs or Muslims in traditional dress. Those that were present were servants and lowly employees of the local café and therefore invisible.

The Americans and Israelis didn't realize that Mustafa Farouk had any English. He listened carefully to the discussion on how to manage a smallpox outbreak while he collected coffee cups and stacked the saucers on a tray, but he only understood about one word in three.

Mustafa had never finished high school—he had two years more than his older brother Ali and one year less than his brother Zawi—but he had street smarts and a very, very well developed survival instinct. How else could he feed his five children, if he didn't seize every opportunity to better himself, to increase his meager cook's salary? His primary job was sous chef on the Oberoi Hotel Boats. When he wasn't working for Oberoi, he filled in at the Café Shalimar near the Oberoi quay. It was handy being on the Shehrayar, since it often moored at Luxor where his family lived.

He glided noiselessly between the clusters of worried officials. Mustafa handed out packets of sugar labeled in Arabic and little napkins, bowing politely to the Western infidels.

Mustafa continued to serve coffee, observing with pleasure the beads of sweat on the foreheads of worried men and women. The threat of a widespread epidemic was enough to throw these Westerners into a panic. His brothers would be so pleased with his report at the meeting tonight.

Mustafa adjusted his robe so he could scratch his belly and smiled at the thought of the bonus he would receive.

Chapter Fifteen

The same morning, Luxor

James collected the rest of his gear from the stateroom he had shared with Lisa.

She sat on the bed watching him. "This sucks, James. I really wish you weren't moving off the boat."

"I don't have a choice, love. Mallory's 'request' felt like a subpoena. I think I can be useful. Besides, I may as well help the medical team instead of lolling around a clinic room by myself." He fastened his suitcase and straightened up.

Lisa stood up and walked into a big bear hug. He didn't kiss her, even though they'd agreed he could hardly be contagious so soon after his exposure to Safaa Farouk.

She looked up into his face, hoping her tone of voice wasn't shrewish. "I suppose Betsy Grover is going with the medical team too?"

"Yeah. She was exposed just as much as Stanley or myself, and she is a trained nurse. Don't worry, I won't let her latch onto me."

Lisa reflected that "latch onto" was not exactly how she'd describe Betsy's heavy flirtation with her husband.

"Hmph," she said, lowering her gaze. "I don't trust her around you."

James gripped her by both shoulders. "You know there's nothing between me and Betsy. You, on the other hand, have this long history with Greg Manzur. I'm the one who should be worried!"

Lisa wriggled away from him. "Don't be ridiculous! Greg is past history."

Now it was James' turn to snort his disbelief. "The way he looks at you says otherwise. Try to keep him at a safe distance."

"Oh, right. And you try to keep Betsy Grover in line!" she retorted as they left the room.

Their parting was on the cool side and altogether too public. Dismally, Lisa watched James walk away from her up the steps onto the quay. She knew she had blown it by bringing up Betsy. Misunderstandings between them were rare and left her feeling vulnerable and unsettled. James' promise to phone her often should have helped, but this time the threat of a particularly nasty and lethal disease hanging over them made the physical separation seem like a dress rehearsal for something worse.

"How about a coffee? I'd like to talk with you," said a familiar voice at her elbow.

"Okay," said Lisa. "I could use a hit of caffeine."

Greg steered Lisa to a busy café that boasted a lime green exterior in peeling paint and strings of beads across the narrow door. Inside, Greg and Lisa sat on ancient metal chairs at a small, rickety table. Around them were mostly men, smoking the nargillah (water pipe) and gossiping. Their inscrutable gazes fastened on Lisa's hair and body, making her uncomfortable. Ignoring both the gazes and the buzz of conversations, Lisa stared at Greg's profile as he lit a cigarette. A straight-bridged nose, thin lips, dark brown hair that drooped enticingly over his eyes. She had always found him attractive; he was more so now. His face had a few lines around the eyes and nostrils that suited his thin build and made him look like an experienced hunter.

Greg's expressive dark eyes regarded her now. He spoke softly in English, his voice almost inaudible against the background of Arabic and the clink of china. "The meeting with the CDC and the Egyptian Ministry of Health was moderately successful; their job is coordinating the medical response to the smallpox outbreak, and that's shaping up. Their real problem is going to be manpower—finding enough reliable people to enforce a quarantine of the Farouk friends and relatives who won't accept the vaccine. My job is the intelligence side, and I have a manpower shortage, too. Since your husband is going to be very busy helping the CDC, you'll have time on your hands. You could help me, if you would, with trying to pinpoint where the terrorists will strike next."

"But what about the Israeli agents you brought with you?"

"They don't look like tourists. You do."

"You mean, for once, being a blonde is an advantage in the Middle East?" Lisa laughed.

Greg smiled at her. "Exactly that. On one level, you don't stick out because everyone will assume you are just another Western tourist."

She hesitated, wondering if "helping" was a good idea. Time spent with Greg was dangerous; too many old feelings came bubbling to the surface. Not to mention that the last time they'd worked together in Israel, Greg had come very close to getting her killed. On the other hand, it might be just the distraction she needed. Her second honeymoon was in ruins. If she didn't keep herself occupied, the worry would gnaw away at her semblance of calm.

"Do you really think I can help?"

"Yes. You haven't spent much time in Egypt, but you have been both an archaeologist and a tourist at sites in other countries of the Middle East. You can help us spot typical tourist behavior, patterns of movement that might be used to plan an attack."

Lisa swirled the coffee in her miniature cup. "You mean," she said slowly as the caffeine kicked in and her mind engaged, "I can see patterns like how tour guides muster their clients in various key locations. Such as how there are always loner tourists who wander away from the main group—usually men—to take photos. That sort of thing?"

Greg's brown eyes regarded her with approval. "Exactly. You've already suggested a difference in how women and men behave on the same tour. Keep going."

"Well, on our tour, the elderly tourists seize every opportunity for standing in the shade or sitting down. It's hard for them to stand still for long stretches. Men and women behave differently. Women are the followers and good listeners, the ones who hang around our guide and ask most of the questions. Men wander. No, wait a minute—male movement depends on nationality! The Japanese guys hang with the group, photographing everything, but American and British guys lurk at the back, smoking or fiddling with their cameras. And guys like James, who've always been independent, wander in and out of rooms, looking for things off the beaten track but still within ear shot of the guide."

"Archaeologists, on the other hand," continued Greg, "take close-ups of mud-brick wall construction or damage to the stone from Napoleon's soldiers sharpening their swords. Or cavities left by metal clamps used to fasten blocks of stone together at the Greek sites." He raised his eyebrows at her knowingly. "And they hardly ever listen to the lecture because they've often given one just like it at home."

"Touché," said Lisa. "We're supposed to go to the Temple of Hatshepsut and the Valleys of the Kings and Queens later today. I can observe not only my own tour group, but others as well. All these sites are jam-packed with Japanese, Germans, and Americans milling about. The other groups won't even notice me if I switch groups occasionally.

"Excellent," said Greg. "That is your first assignment. But Lisa..." His features tightened. "No heroics this time."

"Heroics!" scoffed Lisa. "No fear of that! I have no intention of dangling over a cliff again or being shot at the way we were at the Dead Sea." She eyed him curiously. "But what will you do?"

Greg stared at the glowing end of his cigarette and his eyes narrowed. "Lurk. Lurk near every ticket booth at every site, looking for exchanges of objects, suspicious packages, and odd behavior. My men will help me."

She saw the thin lines of strain around his lips and eyes and a pang of sympathy tinged with something more potent moved her to say, "You're really worried, aren't you? You think this group, whoever they are, will strike a major tourist site in a way that makes a lot of people sick at once."

Greg's gaze traveled from her eyes to her lips and then moved away from her face. "It's a foregone conclusion that they will try again. The incidents so far have been tests, tests to see if the virus works and then to find the best way to transmit the disease. So far, only a couple of people have been infected at each site, but we must assume that the terrorists' ultimate goal is to infect as many people as possible in a short period of time. So that means, when they strike again, it will likely be not just one site, but several simultaneously."

Lisa shivered. What an exit line; it left her with nothing to say.

They finished their coffee, and then Greg walked her back to the cruise ship.

It was hot enough to fry eggs—on the ground. Lisa trudged up the long flight of steps to the Temple of Hatshepsut, admiring the stark scenery on either side. Tan and brown and ivory rock formed fantastic ramparts for the archaeological site. Nothing grew nearby, but Lisa knew that there were oases in the Western Desert as pretty as anything around the Dead Sea in Israel.

Beside her trudged the other tourists from the MISR Travel group. Or rather half of them. Their guide, Mahmoud Sadaat, thought he had found a clever solution to tourist unrest. After listening to continuous complaints about the disruption of the tour and the amount of money people had paid to come to Egypt, he gathered them all on the boat's deck after breakfast.

"Listen, all of you," Mahmoud said. "I have spoken with health officials. They need to interview each one of you about where you have been in Egypt before this tour and which vendors you go to for shopping. But they do not forbid us to go to local sights. So, half of you with last names beginning with A-M, come with me in the bus after lunch to see Temple of Hatshepsut at famous site of Deir al-Bahri. The rest of you with last names beginning with N-Z will stay here on the boat today for interviews. Tomorrow we switch places. Okay?" He smiled, expecting agreement and good-natured compliance with his plan.

It was not okay. Poor Mahmoud.

"I object!" said an elderly Bostonian wearing Bermuda shorts. "I already talked with the Centers for Disease Control guys, I do not have to stay here."

"That was not official interview," said Mahmoud, beginning to sweat.

"I don't see why they have to talk with us," whined Linda Miles. "We weren't anywhere near the woman who got sick. We didn't even go near her house!"

"There are two confirmed cases of smallpox," said Lisa, trying to be helpful. "The other one is a German on the hotel boat that's been following us up the Nile. The medical team has to talk with everyone who might have had contact with the German to prevent further outbreaks. And there's also B—"

She stopped as she remembered that the new information about Bertha had come from Greg, and Linda and Patsy didn't know yet that Bertha was almost certainly the next case of smallpox. Part of her wanted to throw the information in their faces, but she didn't want to cause a panic among the other tourists.

"Oh, you know all this because you're married to a doctor, I suppose?" said Patsy Davis with a nasty little smile. "Well, let me tell you, I don't have to subject myself to this kind of treatment. Just because some of you wandered away from the tour group at Karnak…"

It had been an unpleasant half hour.

Now her group was gathering on the second terrace of the Temple of Hatshepsut, with a fine view of the façade in front and above them. Lisa positioned herself at the rear of the group so she could observe her fellow tourists while Mahmoud gave them an introduction to the site.

"Hatshepsut was one of the few Queens of Egypt. She was Pharaoh for twenty years and is often depicted as a man. If you change her name to English, it is pronounced 'hat-cheap-suit.' Try it," he began.

"Hat, cheap, suit," repeated the group, all except for the male lurkers at the back and the Iowa trio next to Lisa. She could see Patsy Davis struggling to replace her film—in full sunlight, no less—and Nancy Keller trying to assist her. Nancy was the slowest of the bunch. She had a tendency to ask where they were or what they were seeing right after the guide had explained it, and then say "Oh, really?" in a slow drawl that made her sound stupider than she really was. Nancy was also the one who was always late when the group was supposed to get back to the bus because she couldn't tear herself away from the souk.

"She had a son, Tuthmosis III, who ruled with his mother until she died. He was so jealous of his mother's former power that he made his men mutilate every monument that showed his mother. You can see cuts in the stone at many sites," continued Mahmoud.

Nice kid, thought Lisa. But how typical for this part of the world. Rulers all over the Near East—Egyptian, Assyrian, Babylonian—were fond of rubbing out their predecessors in every way. None of them liked the frescoes, stelae, and statues mounted to honor previous rulers. They thought it detracted somehow from their power and influence during their own reigns. This tendency to rewrite or erase history was at least five thousand years old.

Off to her left, David Wallis, the Chicago dentist, was beginning to wander as usual. David had somehow managed to get his interview done early—Lisa suspected he had jumped the queue—so he had come along with the A-M crowd. At each stop, Lisa watched him listen to the lecture for a few minutes before he shrugged his camera up higher on his shoulder and moved away to find something to photograph with his oversized Nikon. No doubt the camera was digital, with huge storage capacity, and all the latest technology. The pockets of David's vest bulged with lenses and other paraphernalia.

In sharp contrast, Barbara Reese carried only a straw purse with water bottle, sunglasses, and cigarettes. She was dressed comfortably in loose linen pants, sandals, and a short-sleeved cotton tunic. Her husband Ron and son Nicky looked equally comfortable in shorts and T-shirts.

Nicky ran up to Lisa carrying a pint-sized digital camera. "Take my picture, okay?"

Lisa agreed and Nicky posed sideways with his hands held flat in front and behind him, like a cartoon of an Egyptian. Barbara laughed and pulled out a cigarette. Lisa frowned as she realized the Reeses should be back on the boat—had they managed to get their interviews done early, like David Wallis, or had they just taken off?

Mahmoud was explaining the layout of the temple now. "We have some of the best preserved colors in all Egypt on the frescoes here, except for the one or two tombs we will see at the Valley of the Queens. If you stop on the front of the building, on your right, you will see some color. Don't forget to look up at ceiling."

"Now," Mahmoud checked his watch, "You have forty minutes. Back to the bus at four p.m., then we will either stop at Colossi of Memnon—these are huge statues built by Amenhotep III—or another site on the way back to the cruise boat."

Nancy spoke up. "Mahmoud, what building is this?"

"The Temple of Hatshepsut, Ms. Keller," said Mahmoud patiently. "We are at Deir al-Bahri, one of the best sites in all Egypt."

"Pay attention, Nancy!" called a young man at the back.

"Yeah, or you'll have to stay after school!" called another.

Sniggers greeted these remarks. Nancy ran a hand through her white, cropped hair and said, "Oh. I thought we were at Luxor. Aren't we?"

"We're still in the Luxor area," replied Lisa with a smile. She liked Nancy better than any of the other Iowa ladies. Patsy was too bitchy and small-minded, and Linda Miles was a constant giggler.

"Oh, thank you, Lisa!" said Nancy with a broad grin that lit up her round face. "I can't hear all that well, you know, and my camera was giving me so much trouble."

"No problem," said Lisa. "Nancy, you might tell Patsy to change her film in the shade next time. She probably ruined that roll by standing out in the full sun." She had no intention of inviting sarcastic comments by telling this to Patsy herself.

"Oh, I didn't know that!" said Nancy. "I thought you could change it anywhere. Jeez, it's hot here. It makes my head just ache."

She moved away from Lisa, rummaging in her enormous flowered bag for her own camera. Today Nancy wore an aqua sleeveless tank, three necklaces, and shorts so short she had better not bend over. Her chest and heavily freckled arms showed signs of sunburn, especially in reddish patches around the top of her shirt.

Lisa strolled over to the colored frescoes Mahmoud had indicated. He was right; the blues and greens were particularly striking, but she had trouble making out the subject of the scene. She opened her Lonely Planet guidebook and found the passage. She was in the Chapel of Anubis, the jackal-headed god who presided over mummification. The co-rulers, mother and son, were shown in the presence of three gods, but the mother's image was disfigured, just as Mahmoud had said. The son had rubbed his mother out, or rather had directed someone to hack away at the image with a chisel. What an ungrateful child, especially after his mother had given him a kingdom.

Lisa continued her tour of the site, hunting down frescoes while observing small groups of fractious tourists. Everyone was hot, disheveled, and vocal. An English couple was the loudest.

"Next time, let's just go to Cornwall as we usually do," said the middle-aged woman wearing a navy blue shorts combo without a hat. A case of sunstroke waiting to happen.

"I'm for that," said her significant other as he mopped his sweaty brow with a dirty Kleenex he'd pulled out of his back pocket. "Any place with a beach and a pub will be all right with me." From the look of the belly that hung over his belt, he consumed a lager or three on a daily basis.

Lisa headed back for the bus, admiring the subtle colors of the rocks around the plateau. Tan, mauve, brown, cream, gray—a painter's dream. And at sunset, when all the warm colors became tinged with blues and purples, the view would be spectacular. She took a photo so she could print it out and try painting it at home. Oil or watercolor? Maybe she should try painting on glass to get the luminous quality of the sky.

As she rolled up the sleeves of her long-sleeved shirt, Lisa noticed the driver standing close to a short Arab dressed in a crumpled tunic over white pants. They were deep in conversation, both smoking hand-rolled cigarettes. Just as she came into earshot, she saw the short man pass the driver a small package wrapped in brown paper and tied with string. The bus driver tucked the package inside his shirt and looked straight at Lisa.

Quickly she pretended she had a pebble in her shoe and leaned over.

But her mind was racing. Just before the package disappeared, she'd seen a little red logo on it. Was it a sphinx? It was the right shape—rectangular with a projection at one end—for a recumbent lion with a human head.

Why did that set off alarm bells in her head?

As she greeted the driver cheerfully and asked for another bottle of water, Lisa tried to remember where she'd seen the red sphinx before.

Chapter Sixteen

That afternoon, Luxor

Greg hadn't told Lisa the full truth about why he was in Egypt. Nor had he decided exactly what Lisa's role would be. He had to strike the right balance: tell her enough to keep her interested and make her think she knew what was going on, but not so much that she could betray one of his agents if she were caught and interrogated by the terrorists. He knew damn well that James Barber would have his hide if Lisa got hurt, but part of his job was to ruthlessly evaluate each contact's usefulness with little regard to his or her personal safety.

At the moment, his goal was to persuade young Abdul Farouk to give him information about his relatives. Greg entered the isolation ward of the field hospital and checked with the nurse. He found the correct room and approached the narrow bed.

He saw a slender youth with the dark, liquid eyes of the Middle East set in a too-thin face now marred with a few ugly pustules. The kid looked sick, but he didn't seem to have as many sores as his aunt Safaa. Perhaps this was a milder case?

Abdul pushed himself to a sitting position, leaning back against two dirty pillows.

Greg looked at him carefully. The boy appeared to be lucid, and he'd heard the fever was gone. "How are you feeling today?" he asked, drawing up a folding chair.

"I am better," croaked Abdul with a slight twist to his lips that might have been a smile.

"Do you know who I am?" asked Greg.

"You work for the American government?" Abdul's English was surprisingly good.

"Not quite. My name is Gregory Manzur. I was born an American, but now I live in Israel and work for the Israeli government. I am here in Egypt to investigate the sickness you have and make sure no one else gets it."

"My aunt Safaa died of it," said Abdul. A tear slid down his thin cheek and mixed with the seepage from one of his sores.

Greg winced at the sight. "I know," he said. He studied the young man, trying to gauge his intelligence. "You work for your uncle Ali Farouk, right? Selling bags to tourists?"

A wary nod from Abdul.

"How would you like to make more money than he can pay you?"

Abdul's eyes widened but otherwise he remained motionless.

The young man knew how to control his facial expressions even if his eyes changed. Greg chose his words carefully. "If you help me, I can pay you well. To start—to show you I am not lying—I will pay you fifty American dollars if you answer my questions truthfully. Okay?"

"Okay." Abdul's tone was noticeably brighter.

"To make sure no one else gets sick, we have to figure out how you got sick. Did you have contact with anyone—your family, friends, anyone—who had a rash?"

"I already answer this question—men from CDC come to my house."

"Yes, I know. I have to ask you again."

"Only my aunt Safaa."

"Has anyone else in your family been sick recently?"

"Just me."

Greg watched Abdul's face carefully as he switched gears. "Have you seen any small packages or any deliveries of goods you were not expecting at your family's house or place of work?"

A flicker of fear crossed Abdul's face. "No," he stammered.

Aha. He was onto something here. "Are you sure? Other peoples' lives depend on our finding out the truth."

"I am sure," said Abdul, but his eyes were huge.

Greg knew he was lying, but now he looked scared and frail. Perhaps Abdul was not yet capable of hiding his feelings, but even that could be turned to Greg's advantage.

"Abdul, this is very, very important." He leaned forward so Abdul would not look away. "Someone in Luxor is trying to make tourists sick deliberately. They are bringing in small bottles, or maybe small packages, of a powder that makes people sick, and spreading it around. Have you seen anything like this?"

Abdul shook his head violently. His pale forehead glistened.

"Your uncles are known in this area for smuggling and other

activities outside the law. Are you sure they have nothing to do with this?"

The young man remained mute.

Greg stared hard, willing Abdul to change his story. Whom was he trying to protect? He was determined to get to the bottom of this, but right now he realized further questioning would be futile. Greg would put more agents around the Avenue of the Sphinxes and make sure everyone who went in and out of the Farouk household was watched.

"If you help us, you can earn a lot of money," he said carefully, watching Abdul's eyes. He extracted a couple of twenties and a ten dollar bill from his wallet and laid them on the thin white blanket

The boy stared at him like a cornered animal. He said nothing, but his left hand shot forward and seized the bills.

"I will return," Greg said, and left the curtained alcove.

It was time to question Abdul's father and uncles.

Lisa pulled her cell phone out of her waist pouch as she mounted the narrow stairs to the boat's deck. The sun was just setting, turning the Nile landscape into a tapestry of green and gold tinged with orange. Clumps of papyrus huddled against the banks where two small boys were bathing nude. Two feluccas glided behind the Shehrayar, tattered brown sails full in the freshening breeze.

Lisa was sweat-stained and exhausted, but she craved solitude more than a tepid shower. Her insides screamed for an icy cold beer. She plunked herself into a deck chair and smiled at the uniformed waiter at the little bar. Immediately, he glided over to take her order.

"One Sakara? Yes, Madame."

"And some snacks, please, if you have them." She knew it wouldn't be the nuts she was used to but rather a spicy mix of lentils and toasted grains mixed with raisins. Lisa sighed in relief, savoring the temporary quiet. Soon, the freshly showered tourists would mob the deck, drinking and gossiping until the dinner bell rang.

A slender minaret on the far bank caught her eye. It made a perfect black silhouette against the orange sunset. " 'Now I know we're not in Kansas, Toto,'" muttered Lisa, as she fiddled with the cell phone in her lap.

The cold beer, a frosty glass, and a little dish of lentil mix appeared magically at her elbow. She crossed her legs, poured the beer, and took a long swig. Her recent email to her best friend, Ellen Perkins, replayed in her mind:

"Hi, Ellen. Here I am on the cruise boat, sitting at their little computer in the lobby. The connection is like molasses, so you may get this in a week—or never. We're in Thebes, or ancient Luxor. But don't envy me because I haven't told you what's happened..."

She sent a brief account of the smallpox breakout, her temporary separation from James, and the awkward reunion with Greg Manzur. Ellen had known Greg when she visited Lisa in Israel, so this news would inspire her to write back with reams of advice for Lisa. And Ellen knew all about being part of a triangle, having dated James just before he met Lisa.

James. Where was James, right this minute? How was he? Each time she called him, she dreaded hearing that he had a fever, or a rash.

She turned on the cell phone and punched in the number of his room.

"Hello?"

"James! I didn't really expect to catch you so easily." She was also surprised that her cell phone worked on the boat. Half the time, she couldn't get reception in Boston, where she lived and the cell phone company was based. "How's it going?"

Lisa heard him groan. She imagined him rubbing his temples and pushing long fingers through his thick curls.

"Okay, I guess. Stanley, Betsy, and I are stuck in one part of the clinic helping with paperwork. We're not allowed to leave the complex until it's certain we're not getting smallpox. The CDC reps and Egyptian medics have been interviewing the relatives of Safaa Farouk. A slow business, even with translators. They are so frightened, not just of the disease, but of any government representatives."

She wanted to ask how much he saw of Betsy but instead said, "Have you found out how Safaa got sick in the first place?"

"No. I think that will take awhile. I had a friend in the EIS once who said it took forever, tracking down every contact and doing interviews. I imagine it's twice as hard over here."

"EIS?"

"Epidemiological Intelligence Service. You remember, that's the outfit that almost recruited me when I was paying back time for medical school, before I became a radiologist. They're part of the Centers for Disease Control."

"Oh," said Lisa, taking another heartening swig of cold beer. "What about you? Are you okay?"

"Lisa," he said, and she heard the smile in his voice. "It's too early for me to show any symptoms. And I may not get it. From what I heard at the briefing, the most contagious stage of smallpox is when the rash first appears. Safaa Farouk was well beyond that point when I examined her. And Bertha Miller—if she does have smallpox—was at an even earlier stage of the disease."

Lisa was not reassured. No matter what James said, he was still in danger of getting one of the most horrible diseases ever known. She hadn't told him about the pictures of smallpox victims her first husband, Tom Henderson, had shown her: the active cases covered with pustules, and the few survivors with pockmarks all over their faces. And that she knew blindness was a common side effect.

She shifted the cell phone to her other ear and changed the subject. "I miss you. And I'm sorry I acted stupid about Betsy Grover."

"That's okay, you had cause. I mean, not on my side, but she's an awful flirt. I miss you, too. I thought we were going to spend this vacation together, not in two different places. I'm in semi-quarantine and you're in a floating hotel."

"Yeah, hardly seems worth the trouble of making all those childcare arrangements and then walking into this!" said Lisa. "Speaking of children, I called Boston yesterday. Sam and Emma are fine. Sam said to tell you the Patriots are having a fantastic season."

James groaned. "And I'm missing it."

They were both silent for a moment.

Lisa's chest felt tight. "James, I'm scared. I could deal with all this so much better if you were with me. I've been having nightmares, too. The kind that make it impossible to go back to sleep."

His voice was gentle. "I know, honey. The whole situation is beyond anything we have ever faced before. Try and distract yourself with the archaeology—that usually totally engrosses you."

"Okay," said Lisa, but tears gathered in her throat and threatened to drown out the taste of her beer.

"Honey?" said James.

"Hmm?"

"How about for our next vacation we go someplace really tame—like New Hampshire?"

Abdul lay on his narrow bed, clutching the money and quivering with reaction. He opened his fist and stared at the three bills. Fifty American dollars was more money than he'd made the past two months. And the American man with the penetrating eyes had given it to him for answering a couple of questions! How much more could Abdul earn if he told the full truth about the shipments? Could he really trust a Westerner?

He was sure Manzur had not believed him. After all, he said he'd be back, and he looked like a man of honor.

But how could he betray his father? It was against tradition, against religion, against all his instincts.

He had an uneasy relationship with Zawi. Nothing Abdul ever did was quite right, even though he was good at schoolwork. When he brought home good grades in biology, his father seemed upset rather than pleased. Nor did he like to hear Abdul talk about leaving in December to study in Cairo.

"I really want to do it, Papa. I would be a good doctor, and I already know I am not afraid of disease or blood."

Zawi Farouk nodded. "It is an admirable goal. A good education will bring you much more money. But we need your local income for at least another year. Perhaps you could start your courses by correspondence and still work for me."

"Papa, to study medicine you cannot do it by correspondence. You must be there for laboratories, for practical work in the hospitals. And Cairo has the best universities."

"Then you must wait," said his father with a scowl.

"But I could live with Rashid and work part time. That way I could send you some money every month."

"Bah! You have no idea how much it costs to live up there! You would have to pay for books and registration fees, and buy some of the food for Rashid's household. There would be nothing left to send home! Besides, look how many young men get degrees and then can't get jobs!"

"But Papa..."

Zawi shook his head. "Forget this idea. Think instead about how to earn more money here in Luxor. You must find something else the tourists like to buy, and you will increase your hours in front of the Karnak ticket booth."

"Papa, I can't work more hours and still finish high school..."

They argued round and round and made no progress. Now the memory made Abdul's head ache. He was still sick; it was so hard to think clearly.

How he longed to get away from this house and work for someone beside his father and uncles. Farzad had once talked about setting up together some day. They'd start a new shop, selling alabaster or maybe carpets like cousin Omar. Abdul could still do that for a little while, but things were changing. Farzad would use his profits to start a family while Abdul used his to go to medical school, the best one he could afford.

Abdul returned to cold reality in the white-sheeted bed and stared at the curtain next to him. Manzur had said he would pay Abdul for more information. He had an idea how the ugly disease had entered his house, but if he said anything, and his uncle Ali and his father Zawi found out... He didn't want to admit, even to himself, what his relatives were capable of doing to anyone who crossed them or countered Mustafa's ideas of Jihad. Betrayal of family was the ultimate sin…

But he couldn't get away, he couldn't start on his own, without money. Abdul desperately needed a source of income that Uncle Ali and Zawi didn't know about.

The dizziness and sweating returned and he could not plan.

He would wait until he was better, then he would decide.

Chapter Seventeen

That evening, Luxor

Lisa leaned over the railing, gazing down at the dark water. She hadn't been able to eat much of the excellent dinner—lamb and couscous, a multi-bean salad with cucumbers, mint, and tomatoes—and had only picked at the delicious almond pastry. Depression made her droop, and she was afraid to return to her room because she knew she'd give in to tears again.

The water trickled past the boat in little waves and rivulets lit by the un-curtained windows below her. Behind Lisa was the low murmur of tourists and the occasional laugh as the after-dinner crowd relaxed and compared notes on their interviews and the archaeological sites. No one sounded worried. At dinner, Lisa sensed a feeling of complacency, that smallpox and other horrors could happen to the natives here in this third world country, but not to them, oh no. They were Westerners. Although the hotel boat contained mostly Americans, there was a sprinkling of Brits and Koreans and even an Indian couple.

Lisa wandered down to the stern of the cruise boat, grateful that no other large vessel was moored next to them to block the spectacular views of life along the Nile.

The rear of the boat was quiet, with no one there except herself. She listened to the lapping of the water and felt cool air moving and freshening.

She tried to imagine James was right beside her. He'd rub her shoulders and let her lean against his broad chest. Some second honeymoon this was turning out to be. She regretted now that they'd both been too tired to make love their first few nights in Egypt. That was one of the things she'd been looking forward to—a little leisurely lovemaking with no chance of Sam or Emma interrupting ("Mom! Can I have a glass of water? "Mom! I had a bad dream…")

But her darling kids were thousands of miles away. Right now, Lisa would give all she had to gather her little family in a safe place—a magic place where James had never been exposed to smallpox, where no sickness or other harm could touch any of them. Where they would be together.

She sighed and her body slumped so that both elbows rested on the railing.

"Misa' al khayr (good evening)," said a soft voice below her.

"Mustafa, misa' an-noor," replied another.

Lisa strained her eyes and was able to make out two men at the end of the gangway that led to the quay.

They moved closer together and stepped into the dim light of one of the lamps near the boat's entrance. The shorter, squatter man was one of the ship's crew—maybe a cook, or a waiter? He sure looked familiar, but Lisa couldn't make out his features.

The taller man pulled the crew member back into the shadow with a jerky, furtive movement. Suddenly alert and a little scared, Lisa withdrew her body into deeper shadow and sank into a crouch.

The shorter man passed a small package to the taller one. Brown paper, tied with string. About the same size and shape as the ones she'd seen before. Here, in Luxor—and oh, my God—at Giza—the other package she'd seen while they were waiting for the ambulance to take Bertha Miller away! It was too dark to see if there was a little red sphinx, but she suspected there was.

And she remembered where she'd seen the red sphinx before: on the wall next to the door of the house on the Avenue of the Sphinxes.

The house where James had visited Safaa Farouk.

Holding her breath now, Lisa prayed for the men to leave so she could dig out her cell phone and call Greg. An agonizing five minutes later, when her knees were about to give out, they did. She rose to her feet, suppressing a groan, and flipped open the phone.

It was dead.

Zawi and Mustafa Farouk had no trouble eluding the man hired by the Egyptian Ministry of Health to guard the front of the Farouk family complex. The man was local, and therefore bribable. All it took was a few extra packs of cigarettes, and they were out of there.

The two brothers took a local taxi back to Mustafa's house, three doors down from where their older brother's wife, Safaa, had caught the smallpox and died of it. This house was not under quarantine since Mustafa had lied to the health officials about where he lived. They entered a long room adorned with luxurious Persian rugs, small brass tables, and embroidered pillows. The men seated themselves on padded benches at one end of the room. Four other men entered the house without knocking and took their places.

The leader, Ali Farouk, looked ten years older than a week ago. His face was drawn with grief and shock and his voice wavered. "The second shipment of disease powder has arrived," he began. "We have tested the first shipment by distributing it in small batches at Giza, Alexandria, and here in Luxor. Now we must decide how to work the simultaneous strike at many sites with multiple victims. But there is a problem. Our suppliers have not told us everything." He passed a shaky hand over his face and reached for his mint tea.

"What do you mean?" asked his younger brother Zawi in a hard voice. He had no respect for Ali, only scorn at his softness, his inability to finish a simple task.

Ali turned his ravaged face towards Zawi and Mustafa. "They did not say how dangerous it is to the carriers. Abdul broke an ampule by mistake, and the dust got in the air—it must be how Safaa got sick. She is dead, poor woman. My Safaa…" His teacup rattled as he set it down, covered his face with his hands, and wept.

"Don't be a fool," snapped Zawi. "Our Hamas contact said it couldn't be transmitted through the air. He knows—they tried it in Israel. Ibrahim says we must put it on something porous, something the tourists will touch—or buy and then touch."

"But your own son, Abdul, also is sick!"

"It is just a fever," said Zawi.

"Our entire family is in danger! I cannot do this…" sobbed Ali, breaking down completely. He rocked to and fro, consumed by his grief.

"Pull yourself together!" said Zawi with a snarl. "Or I will take over! Yes, I will, to save our family's honor and make sure we get paid what we were promised!"

The thin man in the gray turban said softly, "You will not be paid until we have many sick tourists. That is the agreement."

Zawi felt a surge of power as he stood over his weeping brother. His foot twitched as he thought about kicking that tempting backside. Ali was incapable. Ali was a liability—and so was his son, Abdul. He would manage Abdul as he always had.

He stood up and moved away from his grief-stricken brother. "You can count on me," he said proudly to the thin man. "I will do it. That rug vendor did it wrong at Giza. No one is sick yet. We have several options, you know," He smiled. "There are many materials that would work. Wood, cloth, papyrus—these tourists are so very, very gullible. They will buy anything."

The men huddled closer as he told them his plan.

Zara leaned against the wall outside the open window. She was totally bewildered. What were they doing? Making tourists sick? What shipment?

She crouched down into the shadow as the men began to leave the house. Her body shook as she thought of the probable consequences if her father knew she'd been eavesdropping.

Zara didn't understand all that stuff about dust and glass ampules, but she had uncanny recall. She would repeat what she had heard, word for word, to Farzad, and somehow he'd find a way to sneak into the hospital and tell Abdul. He would have to go soon, since she and Farzad were planning their elopement in two days time, three days if the motorbike was not ready. No matter, there were other ways. A carefully worded email from an internet café further south.

She stood up and stretched her cramped muscles, adjusting her black abiyah around her so no unseemly white flesh could be glimpsed in the dimly lit street.

Silently, she faded into the shadows of the narrow alley between Mustafa's house and the neighbor's where no guard was stationed. Only two houses more and she could slip into her own window.

No one, least of all her snoring little niece Ama, would realize she'd been gone from her bed nearly an hour.

Chapter Eighteen

Monday morning, September 13, Luxor

The next morning, Lisa still had not succeeded in talking with Greg. She'd used a phone in the hotel boat's lobby to leave a message at the number he'd given her but he hadn't called back. She had to try again.

The cell phone was useless. Lisa couldn't find her charger, and she had the wrong adapter for Egyptian outlets even if she had found it. Now she'd have to either find the precise new battery that fit in her model of cell phone—unlikely in Luxor, Egypt—or figure out how to make a local call in a café. If she actually connected with Greg, she didn't want the boat crew—or any of her fellow tourists—to overhear her.

Fortunately, Lisa was exempt from further interviews, since the authorities had met with her right after they had talked to James, on the same day Safaa was discovered ill. She had accepted the recommended vaccination after hearing about the possible side effects. She figured headaches and fever were better than the alternative (being covered with pustules or dying). Besides, it was the only way the CDC and the Egyptian health service would allow her to stay on the boat.

The next excursion wasn't until after lunch. Lisa finished her coffee and pushed her chair away from the breakfast table.

"Going somewhere?" asked Patsy Davis. "Today, it's everyone with last names beginning with A through M who have to stay for interviews with the health authorities, remember?" She sounded as if she were enjoying the drama of it all.

"Yes, I remember, but my interview is over. I was one of the first, right after my husband was exposed to the disease."

"Oh, yes, doctors and their families always get special treatment." This was said with a sneer and a shrug.

Lisa suppressed a rude comment. "Where's Nancy this morning?" she asked curiously.

"Oh, she has a touch of food poisoning. We've all been sick on and off since we've been in Egypt. I don't believe Mahmoud anymore when he says the food is safe."

Lisa refrained from telling Patsy what James had said about the food and how to avoid getting sick—Patsy would only make another snide remark about doctors—and left the dining room. She descended the stairs, deciding not to ask the front desk clerk how to find a less public telephone. Act as if she knew where she was going and no one would stop her.

The tourist police stopped her at the end of the gangway. "Where are you going, Miss? Everyone must be accounted for." The young officer was very polite but very firm.

Lisa looked to the right, towards the café where she and Greg had had their little chat. "I just want to buy some cigarettes," she fibbed. "That café over there? I've been in there before, and it will only take a few minutes."

The officer nodded and stood aside. "Come back quickly, if you please."

Before he changed his mind, Lisa strode along the quay, passing yet another mangy stray cat with a torn ear, and pushed through the beaded curtain. She glanced around. Thank God, there was a battered pay phone in a secluded corner. Using a mixture of pantomime and her few words of Arabic, she managed to communicate that she needed tokens. The curly-haired man at the counter took her bills and gave her no change, leaving Lisa with the feeling she'd been fleeced—again—but her errand was too important to waste time arguing. She put in the tokens and dialed Greg's number, praying he'd given her the right prefixes.

"Aiwa (yes)." It sounded like a different person this time.

"Do you speak English?" asked Lisa, clutching the receiver and trying to keep her voice low.

"Yes, how can I help you?" said the voice.

"I am trying to reach Greg Manzur. It's very important."

"Manzur is not here. Leave message?"

Lisa's heart sank into her dusty sandals. "Tell him Lisa Donahue needs to speak with him as soon as possible."

"Yes. I tell him." The phone went dead.

As she replaced the receiver, Lisa felt the curious gazes of several men on her face. She hoped her expression was inscrutable (James would say that was impossible), nodded her thanks at the proprietor, and hurried outside.

Just above the entrance hung a bunch of faded and flimsy belly-dance costumes in pink, green, and what had once been orange. She passed through a sea of rickety iron chairs and small tables in front of the café and walked slowly back towards the Shehrayar.

Now what? Greg knew where to find her, so she must proceed under the assumption that he would catch up with her sooner or later.

Later meant the Valley of the Kings, and no way was Lisa going to miss that. She could scarcely believe she was within a few miles of the tomb entrance where Howard Carter had shone a light into Tut's tomb so many years ago. Carter had been mesmerized at his first glimpse of a rare, un-plundered tomb.

Lisa quickened her pace. Since most of the tourists would be lounging around their rooms or being interviewed until the excursion started, she would grab her notebook and spend a little time on the deck. To hell with writing up stuff for her museum—she was going to play tourist and read up on the tombs.

She could even pretend, for a few minutes, that she was on a normal cruise.

The Valley of the Kings, "Biban el-Muluk," was just as dramatic a setting as Lisa had expected. The movies had not lied this time. Very little grew in this part of the desert except right along the Nile. The bus transported them across the river to the West bank where towering cliffs and piles of tan and beige rock loomed above them, and the air was full of dust. Lisa could feel it creeping into her nostrils and mouth.

She adjusted the plastic string clamp on her Australian, broad-brimmed hat and followed Mahmoud up the first paved ramp. At least she'd brought hiking boots and not flip-flops, unlike so many tourists who didn't appear to worry about scorpions, blisters, or broken ankles. Not to mention fallen arches—touring Egypt required quite a lot of walking, so decent footwear was essential.

Today, as usual, the stragglers of the group were Patsy Davis and Linda Miles, the remaining duo of the original four ladies from Iowa. Nancy Keller had remained on the boat to nurse her funny tummy.

Behind her, Lisa could hear the penetrating voices of Linda and Patsy. Again, they seemed to have forgotten about their sick friend in Cairo.

"Where's my water bottle? I'm sure I packed one today," said Linda.

"You can have some of mine. I really think the tour company should provide bottled water for us—we shouldn't have to pay extra every day," said Patsy. "Have you got my sun cream in there?"

Lisa reflected that MISR Travel's brochures didn't mention most of the "extras" tourists were expected to fork over cash for, like the bottled water and the toilet paper required at every tourist site. Deciding early on to carry her own Kleenex, Lisa usually avoided the outstretched hands of the bathroom attendants. When she ran out of Kleenex, she sometimes didn't pay even then, walking right by the jabbering women and young men who were understandably indignant at not receiving the expected tip. Lisa guessed some of these people eeked out meager livings this way, but a large part of her rebelled at having to pay for toilet paper. After all, they'd already paid a hefty entrance fee by Egyptian standards. Couldn't the site managers furnish toilet paper out of that?

Mahmoud led the group to a covered porch with cement benches and clapped his hands. "Listen to me, Pharaohs!" he cried, using the name the group had elected early on the trip. Having a single, easily recognizable name to yell out made their guide's job easier since he was usually competing with several other noisy tour guides to get his group's attention. "We will have one hour here. I will give you tickets, and you can use your ticket for three different tombs. You must pay extra for King Tut's tomb."

A groan of dismay greeted this information. "Now he tells us," muttered David Wallis, as he fiddled with his oversized camera.

"There is actually very little to see inside Tut'ankamun's tomb," Mahmoud continued. "There is sarcophagus, but no body. All the gold and decorated furniture is in the Cairo Museum—you have seen it already. So, I am recommending some other tombs to you where there is good color preservation." Lisa listened carefully, deciding she was much more interested in well-preserved frescoes. The guide also mentioned early Coptic Christian graffiti that she wanted to see and photograph to show James.

Lisa hoisted her water bottle strap to a more comfortable position on her shoulder and set off for the tomb of Ramses IV. She trudged over the uneven ground, stopping to take a couple of digital shots of the spectacular landscape. One could get a little claustrophobic here, she thought. The looming cliffs and the deeply cut tomb entrances created a distinctly closed-in feeling.

Lisa studied the map of the tomb posted near the entrance so she'd have a firm idea of the layout and not miss any side galleries while inside. Then she lined up to get her ticket punched. Each tourist presented his or her ticket, which was taken by the attendant, punched, and handed back. She hoped the attendant had clean hands (doubtful) and didn't have a cold. What a great way to spread the cold virus.

She stood stock still as her brain repeated her last thought like a stuck CD: a great way to spread a virus. Lisa looked down at the flimsy paper ticket in her hand. It was decorated with a color photo of the Valley of the Kings, labeled in English and Arabic, just like all the tickets they'd used at other archaeological monuments. Paper is porous, said an insistent little voice in her head.

"Hey, are you in line or what?" said a nasal American voice behind her.

Lisa stepped to one side, leaning up against the rock entrance wall of the tomb as her brain seethed with unwelcome scenarios.

What if the terrorists used these tickets to transmit smallpox? Wait, then the Egyptian attendants would be at risk as much as the tourists. But a terrorist group that was bent on killing a lot of people wouldn't care if they infected a few native ticket takers as long as they knocked off a bunch of Westerners. They could take the virus—James had told her it would probably be in a powdered form—shake it up in a bag with the paper tickets, and distribute the tickets to the various sites as usual.

She shivered despite the heat. The heavy air around her seemed to suck the moisture right out of her body. What wrenched her gut now was sheer terror. The ticket method was plausible, it was likely, and her instinct told her it could be the actual means of transmission. She must get this information to Greg.

Moving away from the line of tourists trying to get into the tomb, Lisa pulled out her cell phone again, hoping for a miraculous technological cure that would make the battery work when it hadn't only a few hours before.

After ascertaining that, of course, the phone didn't work, she stumbled over the edge of the paved path, scanning the crowds for a slight figure with dark, floppy hair. No sign of Greg. Probably he would catch up with her at the boat when they returned for lunch.

Hanging onto that hope, Lisa trudged back to the tomb and rejoined the line. She looked at the ticket, already grimy from being clutched in her hand. Did it have smallpox powder on it right now? Her fingers trembled. Maybe, but it was too late now.

What a pity that all her enthusiasm for the Valley of the Kings was quenched by realizing that events were completely out of her control.

Chapter Nineteen

Same day, noon, Luxor

"Collect your bones, gather together your limbs, throw the sand from off your flesh." (Spell 373, Pyramid Texts)

The bus turned away from the Colossi of Memnon on the West Bank of the Nile. Lisa took a last look at the red quartzite monsters that American scientists had proven were carved from stone quarried near Cairo and transported from Lower to Upper Egypt. It was an incredible feat of engineering, especially since the heavy stone statues were hauled against the current, from north to south.

About eighteen meters high, these faceless statues were one of the first things most tourists saw in the Luxor area. What they didn't realize was that the Colossi were once part of a funerary complex larger than Karnak. The sandstone walls decayed much faster than the harder quartzite statues during the annual Nile floods, so little was left to help visitors imagine what the temples had looked like.

Lisa fidgeted in her seat, totally failing to relax because of her impatience to talk with Greg. What if he wasn't at the Shehrayar? Then she'd have to visit the café again. Cigarettes were a good excuse—if Lisa still smoked, she was sure she'd be consuming a pack a day. There were times when she missed having something to do with her hands as well as the calming effects of the smoke in her lungs.

As the bus slowed down to make the turn onto the road running along the boat launch area, Lisa spied a large crowd up ahead. Then, as they got closer, she saw flash bulbs going off.

The international press had arrived in force. Lisa spotted the logos of the BBC and CNN adorning hats and microphones. Lisa remembered James' report of how anxious the Ministry of Tourism was to keep the smallpox "incident" hushed up. Fat chance, he'd said, with several international agencies now involved. And no one had mentioned email. The cruise ship had one public computer in the lobby area that anyone could log onto—for a price, of course—to receive or send email. Probably one of the tourists had written home saying the whole tour was screwed up due to some Egyptian lady getting sick. With a little shock, Lisa remembered she herself had sent such an email.

Then she saw Greg. He was standing near the spot where the bus would disgorge its passengers. Greg looked grim but controlled, intent on the approach of the bus.

Lisa gathered up her camera, water bottle and daypack and descended with the others into a sea of reporters and anxious Egyptian officials.

A young woman wearing opaque sunglasses grabbed her arm. "Miss, what do you think of the latest smallpox case? Are you afraid to go back on the boat?"

Latest smallpox case? Oh, no. "No comment. Let me pass." Lisa shook her head and tried to keep moving, but several other reporters seemed to think she was a good target. Microphones were thrust in her face and one big klutz stepped on her toe.

"Smallpox is a deadly disease," said a tough-looking man with a clipboard. "How do you feel about being exposed to it?"

Behind her, Lisa heard angry shouts and ominous mutterings.

"Are they saying someone is sick on the boat?"

"What's going on?"

"I want to go to my room! What is this crowd doing here, anyhow?"

Using her elbows and daypack as a battering ram, Lisa pushed her way to where Greg stood.

"Lisa," said Greg. "We need to talk." He grabbed her and pulled her away from the throng so they stood in partial shade under the faded belly dance costumes of the kiosk near the boat quay. "One of your fellow tourists on this boat is sick, Lisa. It looks like the first American case of smallpox—in Luxor. We've got about five Farouks in the field clinic already."

Lisa dropped her pack on the ground and rotated her shoulders to loosen the muscles. "Who is sick? I just got here. We haven't been told anything."

"Nancy Keller. Isn't she one of your ladies from Iowa?" said Greg, with the ghost of a smile. She noticed new lines around his eyes from worry and lack of sleep.

"Nancy! I thought she just had tummy trouble," said Lisa, dismayed. "Or at least that's what her roommate, Patsy Davis, said this morning." Naturally, it would be Nancy, the nicest of the four women.

Greg frowned. "Naturally, she paid no attention the other day when the CDC officials asked everyone to report illnesses, no matter how ordinary."

"Well, in her defense, about half of us on the boat have had classic symptoms of food poisoning or bad water. It may be a five-star floating hotel, but I think they wash the fruits and veggies in Nile River water," said Lisa. "Wait a minute, you said 'in Luxor'—does that mean you have bad news about Bertha as well?"

"Unfortunately, yes. Bertha Miller is now the first confirmed American case in Cairo." Greg put a hand on her shoulder. "Hey, Lisa, you need to go get your things right away and come with me. The CDC is about to quarantine this cruise boat. It won't be going anywhere any time soon, and no one will be able to go on and off except already vaccinated health personnel."

Lisa stared at him, her thoughts racing. This was turning into a full-scale disaster. "More smallpox! Oh, poor Bertha. And poor Mahmoud. He was trying so hard to distract us by giving us a good tour."

"Poor Mahmoud, nothing," said Greg. "Because of his inattention to his tourists' symptoms, you may have been exposed to smallpox a second time. But luckily for you, you were vaccinated right after your husband's exposure to Safaa Farouk."

Lisa moved away so he could no longer touch her. "You're just pissed because the limited access to tourists will make your investigation much harder. But how come I'm exempt from the quarantine? Can you really do that?"

Greg's smile was rueful. "Yup. Our investigation is going to be incredibly difficult, and we need every extra set of eyes I can find. Most of the CDC folk will question everyone all over again to try to figure out how Ms. Keller got the smallpox. My team—and that includes you—will redouble our efforts to halt the next attack and prevent new smallpox cases."

"You mean, Nancy didn't get it from someone here in Luxor?"

"I haven't heard anything yet that puts her anywhere near the Farouk family."

"I think she and the other Bimbos came in early. Seems to me Nancy was yakking away about a few days in Cairo before our tour started."

"Yes, we'd heard that," said Greg. His brown eyes focused on something only he could see. "It probably means there was a second incident in Cairo—a second release of the virus at about the same time. If someone else got sick, we should have heard about it by now. But we don't know who released the stuff or how, and we've got to find out, pronto." He flipped his hair back from his forehead. "Go get your things. If anyone stops you, show them my card." He pulled out a business card and scribbled something in Arabic on the back. "We'll put you in my hotel—your own room, of course. But you'll be housed with the other agents and reporters who haven't been exposed directly."

"That'll make James real happy," said Lisa.

She left him and began to push her way back through the throng. Her heart sank as she thought of James' probable reaction. He'd probably think Greg was taking unfair advantage of the situation and that he shouldn't be able to snag Lisa as an assistant Mossad agent. Not to mention that he, Greg, would spend more time with Lisa than James could.

Lisa tightened her grip on her possessions as sweaty tourists jostled around her. James would have to understand. Working with Greg would keep her sane—there was nothing worse than inactivity when she was so worried. And her interest in archaeology wasn't enough to block out the images of people she knew dying in agony or going blind. How could anyone, regardless of religious belief, unleash something so awful on his own people as well as Western tourists?

A healthy surge of anger warmed her. Hell, she wanted to stop the terrorists as much as Greg did. And it would certainly be a lot more interesting than being trapped on a cruise boat with close to ninety upset, scared tourists.

Lisa found herself in a quagmire of reporters and tourists, all vocal and definitely pissed off.

"I demand an explanation!" shouted David Wallis.

"Where is my roommate? What have you done with her?" cried Patsy Davis. "She can't have smallpox, it was the food on the boat that made her sick!"

"If we're quarantined, who's going to pay for the rest of the trip? That's what I'd like to know!" said Wallis again.

"Yeah, who pays? It's a damned expensive 'vacation,'" said another tourist.

The reporters lapped up the confusion and anger, scribbling busily as their photographers and video guys collected enough images for a dozen newspapers.

Mahmoud Sadaat did his best. "Everyone, go on the boat! I will explain everything! Go on, please, up to the passenger deck and we will meet there."

With the help of the Tourist Police, who weeded out the reporters and shoved them rudely behind a makeshift tape barrier, the American group was herded forward and up the stairs to the deck. Everyone crowded around Mahmoud who was now flanked by two CDC officers.

"Ladies and gentlemen," shouted the senior officer. "My name is Michael Browning. I am very sorry to cancel your tour, but we have no choice. Smallpox is a deadly disease. There is no cure. We don't know yet which ones of you have been exposed, but everyone will be vaccinated. Then, you will be isolated here on this boat until we are sure no one else has symptoms."

"How long will that be?" growled David Wallis, who for once made no move to take a picture.

"Two weeks," said the officer, and stepped back at the roar of anger.

"You've got to be kidding!"

"Two weeks! I have to be back at work in eight days!"

"My children need me! I won't have a babysitter after next Saturday!"

The crowd surged forward like a rugby scrum.

The officer raised his voice and one hand in a placating gesture. "You do not understand. This is an international emergency. Smallpox is one of most dangerous diseases in the history of man. We cannot, absolutely cannot, let it spread."

Finally the tourists stopped jostling and began paying attention.

"But we were all vaccinated as kids, weren't we?" said one fifty-something woman with curly dark hair.

"Most of you—the ones older than twenty-five or thirty—were. But the vaccination you had as a child no longer protects you," said the officer. "And younger people never received the vaccine at all after the 1970s."

"How could this happen?" said a tall man wearing shorts and a polo shirt. "I thought smallpox was eradicated long ago. And that the U.S. and Russia signed a treaty not to make any more bioweapons."

"Bioweapons?" said someone else. "You mean this sickness was not accidental? Holy shit!"

"Oh, my God! It's like the anthrax scare all over again!" said woman from New Jersey.

"Or that subway incident—you know, when someone threw nerve gas into the Tokyo subway," said her companion.

Suddenly everyone was talking at once.

"People, people, calm down!" shouted the CDC officer. "We don't know exactly how this outbreak occurred yet, so don't leap to conclusions."

Lisa shook her head. Browning was trying to play down the bioterrorism angle, but rumors had been flying for days now. It was too late. The presence of the press would quickly destroy the illusions of anyone who still thought he was safe. She did not envy Browning his job of trying to enforce quarantine and quell panic at the same time.

As the briefing continued, the tourists became less quarrelsome and more scared. The ruckus heated up again when Wallis raised the question of who would pay for the missed days of touring. Lisa took the opportunity to slip down the stairs to the second deck where her room was. Unlocking the door, she dragged out her suitcase from under the bed and began flinging her belongings in it.

The view from the double picture window made her pause after emptying only one drawer. On this side of the boat, facing away from the quay, there was no hint of the chaos she'd left upstairs. The Nile flowed quietly by the picture window, adorned with clumps of water hyacinth and bits of papyrus leaves. On the far bank, she could see a group of children bathing and animals drinking from the bank. No hint of modernity showed; from what she could see through the glass it might be five thousand years ago.

A door slammed on her level, forcing Lisa to return to her unwelcome task. She collected all her toiletries from the bathroom, already feeling nostalgia for the floating hotel room James had shared with her so briefly. Packing up meant admitting that the second honeymoon was kaput, finished, torpedoed by events beyond her control.

She'd call James just as soon as she was settled in her new hotel.

No, she wouldn't. A wave of shock paralyzed her as she realized the confusion at the quay had driven everything else out of her mind; she'd forgotten to tell Greg about the tickets at the Valley of the Kings and how they might be used to transmit smallpox. That was her first priority.

Chapter Twenty

Same afternoon, Luxor

"Hey, Barber," said the youngest of the CDC team, a compactly built man in his late twenties.

"What now, Mark?" said James, rubbing his itchy beard. Mark Leonard was his minder, the medic who'd been assigned to him to check on James, Betsy, and Stanley every day and give them paperwork to keep them occupied in semi-quarantine—a suite of small offices dubbed the "watch and wait" section of the field clinic. If any one of the trio developed smallpox symptoms, he'd be slapped into isolation and full quarantine.

James had hardly slept the night before. Nightmares—especially about surviving smallpox but going blind—dogged him, so he'd spent hours in front of a computer, reviewing CDC reports that were continuously logged in as new smallpox cases were isolated and new personnel arrived. His clothes were limp, his hair wild, and his mind nearly mush.

"Your wife's tour group has just been quarantined. One of the women—from Iowa, I think—has developed the rash."

"Oh, hell!" said James.

He shuddered as he remembered the tiny pustule he'd discovered just that morning on his forehead. That was a natural reaction to the live virus vaccine, wasn't it? What about his headaches and the chills he'd had last night? He might be getting the disease after all, because it took several days for the vaccine to really take hold. James hadn't been vaccinated until late in the evening of the day after his exposure to Safaa Farouk. He should really ask Mark or one of the other docs to check him over.

He swallowed. At least he needn't worry about Lisa; she'd taken the vaccine at the same time he had, and she couldn't get it from him because he'd been asymptomatic—then. And she'd slept alone since the exposure. But a full-scale quarantine of a cruise ship meant triple workload for everyone else, and even one smallpox case among the tourist group would probably delay everyone's return home.

"She might be exempt, though," continued Mark. "That Mossad agent said he was going to give her something to do."

"Manzur, again!" James slapped his hand on the table. A wave of anger rolled over him. Who did the guy think he was, commandeering Lisa? She had no training as an agent. What could she do, except get in the way and maybe get herself killed? Not to mention the fact that Lisa belonged with her husband—her non-contagious husband, that is. Another thought struck him: even the vaccine wouldn't protect her completely if someone else on the boat had been incubating the disease.

James shook himself. It was easier to think about his wife being exposed to the dangers of a stray bullet or a knife than the horrific potential of this disease. He'd call Lisa and find out what the hell was going on. He wished they could meet, stand outside with about a hundred feet between them, but that was not allowed. James erupted in a sharp laugh.

Mark put a hand on his back. "Calm down, chum," he said. "I know you miss your wife. I'm sure she'll call you later today."

James was not comforted. He knew his wife too well. Lisa would get involved, whether Greg asked her to or not. She had a tendency to walk into danger without realizing it. Like the cats she loved so much, she was curious by nature—and fearless. She'd stick her long nose into every corner of Karnak or that other site—Deir al Something—until someone decided she was a threat.

He stood up and massaged his aching temples. His hands felt clammy on his face so he wiped them off on his trousers.

"Hey, man, are you feeling okay? You don't look so hot," said Mark. He stepped closer, peering at James' face.

"Yeah, I'm okay—I think," said James. "I had another bad night."

"Wait a minute, you've got a pustule. You need blood and swab tests, pronto," said Mark. He left the room to get the sampling kit.

James fought a sudden wave of dizziness as he looked with loathing around the small command center. It had one small window that let in hardly any light, partly because of the file cabinet shoved up against it.

He sat down while he waited for Mark's return. His head spun with all the contradictory information he'd been reading on Soviet stockpiles and what officials thought they knew about bioterrorism.

The most aggravating reports were the ones about bioweapons inspections in Iraq in the 1990s. Hard evidence seemed to slip away into the sand, just like the little bioreactors on wheels the inspectors claimed to have found.

There was one little tidbit James had found that Manzur would find interesting. A Russian microbiologist had connections with Hamas in Israel. Hmm. Was Hamas the conduit for the virus? Everyone knew how porous the border between Israel and Egypt was when no Israeli soldiers were watching. A Hamas connection was a starting point for investigation, but the real questions were who was providing the funding, and who was running the show here in Egypt.

Lisa had given James Greg's cell phone number. James would kill two birds with one phone call. He'd discuss the Hamas tidbit and then turn the conversation to Lisa's role. Dimly aware that he was dizzy again, James dialed from an ancient rotary phone on the desk.

"This is Manzur's cell phone. I am not here right now; leave a message at this number."

Damn the man! Not available to talk in person when you wanted him, and altogether too present when you didn't!

James clutched his head and slumped over the desk.

Greg paced back and forth and waited for Lisa just beyond the crowd of reporters surging around the quay.

He was more anxious than he wanted to admit to anyone. His agents still had no idea how the smallpox was transmitted. How had Safaa Farouk and the German tourist acquired the virus? Where and how had Bertha Miller contracted it? Had they all gotten sick from powdered smallpox, like the scab-covered blankets used by European explorers to infect American Indians? The shadowy group they were pursuing presumably had even less information than the Egyptian government about what worked and what didn't. The terrorists might infect each other while trying to sicken large numbers of Western tourists. A couple of Greg's men were now deployed in towns up and down the Nile, watching and listening for news of fevers and rashes in native families.

And what, exactly, were the terrorists planning to do next? He had no doubt that the main event involved a large number of tourists.

Greg felt in his pocket for cigarettes and lit one, resuming his pacing.

Would they use an aerosol, such as the perfume dispensers they had tried at Megiddo? Or something subtler, like a substance that the tourists had to touch? Powder smeared on baggage, water bottles, airline tickets, brochures—there were so many possibilities. One could argue that the Egyptians in charge wouldn't take chances on infecting their own people, but Greg knew from experience that terrorists didn't care about the native population. They wanted publicity and they wanted dead bodies, the more the better.

They had checked out the Farouk family. So far, no one had talked and Greg and his agents had no solid evidence on any of them. He was torn between wanting to lock them all up so they couldn't do anything and watching them so they'd lead him to the big fish—the international organizers.

Greg threw down the half-smoked cigarette and ground out the butt savagely with his running shoe.

Just when Greg was reaching for another cigarette, Lisa appeared, lugging her small suitcase and a bulging carryall down the hotel boat's ramp. She showed his card to the policeman at the end of the ramp and he waved her through. He watched her walk towards him, welcoming the distraction from his uncomfortable thoughts.

He noticed a gloss on her that was more than youth and good health. Satisfied maturity? Perhaps. Eight years of living, a happy marriage, and success in her chosen career as a museum curator. He had always known that Lisa would never be just a housewife.

Greg grinned ruefully. He had never married, precisely because he couldn't find a woman who would put up with his unreliable schedule. Not to mention his intense need for freedom of movement, no questions asked.

"Come on, Lisa. I have a car for us." He took her bag.

"Okay. Greg, I've got to talk with you."

"Later. Not in front of the driver; he speaks three languages," said Greg as he opened the door for her. "We'll get you settled and then find a drink in a smoky bar where everyone else is too busy making noise to bother listening to us."

The car nosed along the road, parting the reporters and cameramen like Moses crossing the Red Sea, and then they were moving. Quickly, the driver picked up speed as he drove away from the crowded quay with its seedy kiosks, crowded cafés, double-parked cars and mule-drawn carts.

The Hotel Luxor was a modest, mid-sized hotel built of peach-colored stone. It was suitable in every way for Greg's needs since it was centrally located and had satellite reception that worked some of the time. The main building was set back from the street, with a garden and several palm trees in front. Behind the wall surrounding the pool stretched the desert in all its tawny glory.

The lobby bustled with agents, press reporters and photographers from several countries, and CDC and WHO personnel. Greg elbowed his way to the front desk and spoke quickly in Arabic to the clerk.

"He's given you a room on my corridor," said Greg, grabbing the key and Lisa's rollaway suitcase. He strode towards the elevator, his slight muscular body sliding easily between clots of people standing around and jabbering in several languages.

"Wait! Greg, don't I need to register? Show my passport?" Lisa said.

"Later. For the moment, you're with my team and that's enough." One of the advantages of his job was being able to cut through certain kinds of red tape. Not, of course, the sort of red tape he needed for this assignment in a foreign country, but certainly enough to get Lisa a hotel room with a minimum of fuss.

Greg unlocked Lisa's door and handed her the key. "Meet me down at the bar in ten minutes," he said.

"Why don't we talk here? I have a lot to tell you," she said.

"The walls are incredibly thin here, as I've reason to know," replied Greg. "And some of them are bugged. We're safer talking in the midst of people intent on their own conversations. Don't be long, Lisa."

Lisa closed the door, barely resisting the desire to slam it in Greg's face. She was exhausted, filthy, and highly suspicious of Greg's high-handed tactics. She'd jolly well take a few extra minutes to wash her face and call James.

She rummaged in her carry-on for the washcloth she always carried; foreign hotels often didn't provide them. Entering the bathroom and seeing the inviting, pristine shower, she changed her mind. Greg could wait; hot water beckoned. She stripped with the speed of an actress doing a costume change and stood under the water until her skin felt almost sandless and boiled to a satisfactory pink.

Wrapping a towel around her long hair, she crossed to the desk and picked up the telephone.

"Hello. Barber here."

"James!"

"Where are you?" was the rather abrupt reply.

"At the Hotel Luxor. We have a case of smallpox on the boat, and Greg pulled me off..."

"I heard about that. What does he think he's doing? Turning you into a secret agent?" By the harshness of his voice and the slight rise of pitch at the end of his question, Lisa could tell he was upset and trying not to sound jealous. She clutched the phone with one hand while the other rubbed her hair with the towel.

"Relax. He's just using me as an extra pair of eyes, that's all. My assignments so far have been to continue doing the touristy thing, taking photos and gawking at frescoes, and also observing other tourists' behavior." And looking for odd activities among the native workers of the entire Luxor area, but James didn't need to hear that.

James latched onto the part he understood best. "Using you is right. He's a ruthless character, don't forget. He may have been in love with you once but now he's just grabbing you because he's short-staffed. What if the terrorists notice you snooping around?"

Lisa smiled into the phone. James was just making noises because he was worried about her. "I'm not doing anything unusual. I'm strolling around like an ordinary tourist, getting on and off the bus like a good little girl. I'm even buying stuff at the souks I don't want, just to keep up appearances."

James groaned at this. "Flashing your Visa card and your American identity, I suppose? Sometimes, I think being an American over here means you have a target painted on your back. Well, just make sure he doesn't put you in a situation you can't get out of."

"I'll be careful, I promise. So tell me, how are you?"

"This place sucks. And I don't feel so hot. Mark thinks I should—oh hell, I'm probably just short on sleep."

"If you have any symptoms at all, you should get tested!" said Lisa sharply, sensing that he was holding something back.

James sighed. "Don't worry, folks here keep checking me. You're better, right? No fever? No recurrence of the Ramses Runs?"

"It was the Thutmoses Trots. Really, I'm fine."

They talked for a few more minutes and then Lisa flung on some wrinkled but clean clothes and hustled down to the bar.

Greg Manzur sipped his cold Sakara and watched Lisa wind through the clumps of press to the tiny table he'd nabbed in a back corner. Her long, dark blond hair was damp, and she looked more relaxed than she had twenty minutes ago.

"So you connected with James?" he asked, knowing the answer already.

"Yes," she said, pulling out a chair and sliding gracefully into it barely six inches from his right knee. "He's not too keen on my playing assistant agent, but he's relieved that I'm off the quarantined boat." She kept her voice low and directed at Greg.

"You said you had something to tell me," he reminded her, signaling to the waiter. "Let's order first."

While a waiter arranged cold beer and snacks on the round table, Greg studied her face, completing his mental catalogue of how she'd changed. Fifteen years and a few fine lines around eyes and mouth had improved her looks. At close range, Greg could observe how time had broadened her face slightly, softening her formerly sharp features and bringing everything into pleasing proportion. Love and loss, both private and professional, had done the rest. She didn't look like a wide-eyed kid anymore, but a seasoned, compassionate woman. Lisa Donahue knew what she wanted out of life but wouldn't trample others to get it.

The waiter departed. "Okay," he said, "Out with it."

Lisa told Greg about the ticket exchange at the Valley of the Kings and lightweight paper that was used for them. "Don't you think that would be a good way to infect tourists?" she said. "I know some of the native ticket takers would be infected too, but I bet terrorists don't care about that."

"It's definitely a strong possibility," Greg agreed. "As long as the ticket takers and smallpox vendors are unrelated. But if the terrorists have relatives working at the archaeological sites, they might choose another method." He ran through the agent assignments in his head. They were really stretched thin already, but he thought Mark Lewis from the CIA might help with deployment.

"Good work, Lisa," he said, smiling at her. "That's just the sort of thing I hoped you would notice."

Now Lisa was staring off into space above Greg's head. "I wonder...if porous material is used for transmission of powdered smallpox, could they do it through the tourist stalls? Maybe choose one kind of merchandise, say, embroidered white shirts, and infect people that way? That has the added possibility of sending smallpox overseas in the form of gifts." Lisa shuddered. "I assume the virus would survive a transatlantic journey?"

"Probably. And there are way too many good European air connections out of Egypt. We're already combing the stalls around Karnak, the Luxor Temple, and the other sites for such possibilities."

"On second thought, it doesn't have to be the souks," continued Lisa. "You know how the guides take tourists to factories? Our group has been to a perfumery, an alabaster joint, a papyrus factory, and a carpet school. I'd rule out the alabaster statues and the perfume bottles, but what about the papyrus sheets and carpets?"

Greg nodded, thinking Lisa would make an excellent agent but knowing she would never agree to do that sort of work. "Yup. We're onto that, too. But the number of possibilities is staggering—that's why I need you. Try to make James understand that. I don't want him punching me out for kidnapping his bride. And you can slip in some more time on your sphinx research."

At the word "sphinx," Lisa snapped to attention. "Oh my God, I forgot the other thing. Greg, I saw a package with a little red sphinx on it..."

His brow furrowed as he listened to the story of how Lisa had seen a package exchange near the boat the night before the quarantine began. "Why the hell didn't you tell me sooner?"

"I tried to call you on my cell phone that very night, but the phone was dead. Then I went to a café the next morning to call you from there. You were out, but I left a message with someone who seemed to know you. Then, when I finally caught up with you after our trip to the Valley of the Kings, there was the press scrimmage and quarantine."

Greg's thunderous expression cleared. "Communication in this country is abysmal! Never mind, what you saw fits with information that's been trickling in. Do you think you'd recognize either of those men again?"

"Maybe. The light was tricky. I could see the package, but not the mens' faces. The shorter one looked familiar to me, though. I'm pretty sure he's part of the boat crew."

A tall man, thirty-something, with dark hair and hooded eyes, signaled to Greg from the other side of the café. "Excuse me a minute," he said to Lisa and walked over to speak with his agent.

"Bad news," said Rafi softly in Hebrew. "We just fished a body out of the Sacred Lake at Karnak."

"Whose?" asked Greg, although he had a pretty good idea.

Rafi told him. Greg gave him new instructions and then brought Rafi back to Lisa. He introduced Rafi as a "colleague," and Lisa smiled.

"What's up?" she asked, scanning Greg's face.

"The oldest brother of the Farouk family, one Ali Farouk, has just been found drowned in the Sacred Lake at Karnak."

"Natural death?"

He eyed her. "What do you think?"

"So the Farouks are connected with the smallpox plot?" asked Lisa.

"We still have no proof of that," said Greg. "The three brothers are known for petty smuggling, but they've never been involved in anything like this before."

"Maybe they've branched out," said Lisa.

Chapter Twenty-one

Tuesday morning, September 14, Luxor

It was pure luxury to sleep late—or rather, stay in bed late without being subjected to an early wake-up call by Mahmoud to meet the tour group's packed schedule. And she'd certainly had enough of hanging around bazaars and shops while her fellow tourists flashed their Visa cards and tried on unsuitable garments.

She was still lying in bed at nine o'clock, imagining the taste of a Starbuck's latte, when the phone rang.

"Lisa, it's me."

"James! You sound different. Is something wrong?"

"You could say that." She heard him sigh gustily. "I have a fever—and a rash."

"Oh, no! Where?"

"On forehead—a little pustule. It's still early in the incubation period."

"But you think you have smallpox."

"Yeah. Chances are good that I do. But the other doctors told me they won't be able to tell for sure for several days."

"Small comfort. How do you feel?"

"Crummy. Like a bad case of flu plus poison ivy."

"Oh, lovely!"

"And there's more bad news. Stanley and Betsy have it, too. Stanley has a heart condition, so it's more serious for him. We've all been moved from the 'watch and wait' section of the clinic to the Isolation Ward. I can talk on the phone but I can't go anywhere—not unless I'm put in a protective suit like the docs and nurses."

By the time their conversation was over, Lisa was wide-awake and scared. She paced the room, imaging horrible scenarios. James dying. Flying home alone and then having to tell the kids. James surviving, but scarred, blinded, and unemployed. Contracting smallpox herself and taking it home to Emma and Sam…

This wouldn't do, she was driving herself crazy. She decided to use the morning to check on email—maybe there was a message from Ellen—and then try to locate Greg.

After a quick shower, she dressed and headed down to the internet café just next to the hotel. She paid for a coffee (thick as mud and nothing like a latte) and a sticky roll plus a half hour of internet time, reflecting that the proprietor couldn't care much about the safety of his computers if everyone was allowed to eat and drink right next to the keyboards.

When she sat down at a free terminal, her suspicions were confirmed. The computer was an elderly IBM and the keyboard was grimy from many fingers and at least one spilled beverage. Grimacing a little, Lisa logged on and then sipped her coffee while she waited for the slow connection to display her email.

Aha. Ellen had replied. She clicked on the message.

"Zowie, Lisa, you sure know how to pick a vacation! How did you manage to get bioterrorism plus a rendezvous with an old lover? Good grief, I bet James is seething at being cooped up while you're off detecting with Greg. All I can say is, you had better keep me posted. All the gory details, mind. And now, for your spiritual refreshment, a little museum gossip. You won't believe what our Director has done now."

Lisa grinned at the typically irreverent Ellen and skimmed through the rest of the message. She hesitated, not wanting to put her own fears about James into words. It didn't really matter what she said, Ellen would read between the lines and worry with her.

She started typing a reply: "Well, things have changed—for the worse. Now James is sick. Hopefully, he won't get more than a mild case since he's had the vaccine. I don't want to even think about the alternative—smallpox is a really scary disease. As for the rest, you know darn well I didn't plan this situation. I hate love triangles, and you're absolutely right that James and Greg didn't hit it off. In fact, I…"A shadow fell on her screen.

She looked up and found Greg behind her. Quickly, she clicked on the next message on her inbox so the screen would change. Greg's wide grin told her he'd already seen the message to Ellen. "I wish you wouldn't sneak up on me like that," she said crossly.

"Sorry," he said, without a trace of sincerity. "I came to find you because I have to go on board the cruise boat and thought you might like to come. Here, let's get out of earshot of the other customers."

Greg pulled her over to a quiet corner.

Lisa stared at him. “But isn’t that dangerous? I mean, with people on board who are probably contagious with smallpox?”

“Not if we wear spacesuits,” he said, referring to the white, all-enveloping biohazard gear provided by the CDC. “And we’ve both been vaccinated. I could use another pair of eyes. And maybe you’ll recognize the crew member you saw on the boat quay—the one with the package.”

Lisa frowned. “But the passengers and crew will recognize me.”

“I don’t think so. Not in that get-up.”

“Well, okay. What are we looking for?”

“I’ll tell you on the way. I have a car outside.”

Lisa abandoned her computer, reflecting that she’d used most of her paid time already. On her way out, she snatched up her coffee and the rest of her breakfast.

The car wove in and out of traffic, passing donkey carts jostling with every sort of motor vehicle and streams of shoppers who showed no fear at all of collisions.

“So tell me what we are looking for on the boat,” said Lisa.

Greg stared straight ahead, his eyes apparently focused on a pick-up truck full of tomatoes. “Someone, maybe several people, in the crew of the Shehrayar, are working with the terrorists. We’ve had some intelligence that indicates the boat may be a storage facility for packages of the weaponized virus. The packages so far have been quite small—we think that was for the testing phase—but somewhere they’ve got to store larger amounts if they’re planning a bigger attack. They might also be stockpiling guns in case things get out of hand.

“So how exactly do you expect to confirm this?” asked Lisa, grabbing the back of the seat in front of her as their driver braked suddenly to avoid two women in abiyahs and black headscarves.

“We’ll conduct a search of the kitchen and staff quarters, talking to everyone about the disease. We’re going with the CDC team, who will focus on how to prepare the passenger meals without sharing the smallpox. Preventing more smallpox cases is a primary concern, of course, but you and I will also be looking for anything out of the ordinary.”

Lisa tried to imagine how she’d handle isolating passengers while still feeding them and failed. What if one of the cooks or waiters had the disease but had no symptoms yet?

The car screeched to a halt. They had arrived at the CDC field hospital.

Lisa followed Greg over to a truck loaded with biohazard gear and was introduced to a Dr. Mallory.

"You look about this size," said the doctor, tossing her a package containing a white suit, hood, and HEPA filter. "And here's the air pump to attach to your belt."

"Are these suits really necessary?" Lisa asked, thinking how hot and sweaty they were going to be.

"You were vaccinated already, right?" said Dr. Mallory.

She nodded.

"Well, it takes several days for the vaccine to take hold. Until then, you're still vulnerable. And several of my people haven't been vaccinated yet because we're waiting for another shipment from the U.S. We used up the first batch on the index case's family and all those immediately exposed to the disease."

Lisa unwrapped her suit and tried to figure out how to put it on.

All around her, the team members were morphing into abominable snowmen as they donned the headgear and HEPA filters which hid their identities. Greg had been right: no one on the cruise ship would even know what sex she was, let alone who she was.

"Pile on, everyone," hollered a guy with a clipboard.

Already uncomfortable, Lisa joined Greg on a narrow seat and tried to see as the truck sped along the road to the Oberoi Hotel quay.

Soon they were crossing the ramp to the Shehrayar, which was eerily quiet. As the team regrouped in the lobby, Greg summarized the quarantine procedures for Lisa. The medical team's first move, after Nancy Keller had been removed to the hospital, had been to isolate family groups in their rooms since there was no way of telling how many people had been exposed. CDC doctors visited twice per day, looking for rashes or fevers so they could remove active cases to the field hospital immediately.

"Four of the guys with us will be visiting each cabin to check people," he said. "Mallory, Brown, and Wilkes are talking to the kitchen crew, and you and I will tag along."

Lisa nodded. Mallory turned right out of the lobby, heading for the stern of the boat where the kitchen was. They climbed the carpeted staircase to the fourth deck in order to cross through the empty passenger dining room.

At the top of the stairs, they met a passenger Lisa recognized, a young woman. It was Barb Reese, the mother of Nicky. Purple shadows marred the skin under her eyes, which were red-rimmed.

"Excuse me," she said to Dr. Mallory. "I was looking for a waiter to get some bottled water for my son. He's not feeling well, and I think he's running a temperature."

Mallory diverted one of his colleagues to go with her. "I'll send someone from the kitchen to your cabin with water, ma'am, but you must stay inside your suite. Dr. Brown will take a look at your son."

Lisa felt her heart plummet, knowing what Nicky's fever could be—the first symptom of another smallpox case. She hoped fervently that it was a childhood bug instead, or King Tut's Revenge, anything but smallpox. She didn't want to imagine that engaging little face pitted with pox scars.

The group reached the kitchen, which was crowded with waiters milling around while the cooks prepared lunch for the tourists.

Dr. Mallory spoke to the head chef, who clapped his hands for attention.

"This is what we do," said the tall man wearing a white smock. "You knock on the door, pass the tray to the person who comes to the door, and do not linger. You will wear surgical masks while you do this, and you will wear gloves at all times. Dr. Mallory has brought the gloves and masks for you to wear." He opened the box on the floor next to him and began distributing the items.

Greg drew Lisa towards the pantry area and whispered with difficulty through his headgear. "Look through the food supplies. Keep your eyes peeled for anything unusual—packages, small vials, cardboard boxes with no labels—anything that might not be food."

Lisa obediently began opening cupboards and checking shelves. She saw bags of rice and sugar, stacks of onions and potatoes, plastic bags full of pita bread, cans of soup. Most of the labels were in Arabic, but the contents were recognizable from the pictures on the wrappers.

She reached the last cupboard on her side of the pantry and turned back towards the action in the kitchen. An argument in Arabic had broken out between two of the cooks. As the shorter one gesticulated, he tilted his head towards Lisa and she recognized the set of his ears and the way his hair grew low upon his neck. He was the cook who had come out of the kitchen to greet the tourists at their first dinner on board the cruise ship, and he was the same man she'd seen on the quay with the package. She looked for Greg, but he had vanished. Now what was she supposed to do?

The older cook bade the younger one to do something, and he turned away. Lisa watched him go over to a large metal freezer and open the door.

A freezer. Her brain kicked in and her pulse sped up. Hadn't Stanley Rosenberg said smallpox virus could survive being frozen? What if the packages were brought on board and stuffed into the back of the huge freezer? A hotel boat that traveled up and down the Nile would be a dandy storage facility for the terrorists.

She edged around closer, watching as the short cook rummaged around inside the lower part of the stainless steel monster. Out of the corner of her eye, she saw a suited man about the right size for Greg and motioned towards him. He walked over and Lisa sighed with relief as she recognized him through the face shield.

"That man," she whispered to him. "He's the one I saw the other night when I was on deck. And what about a thorough search of all the frozen food?"

Greg nodded and patted her on the back through her suit. "Good work, Lisa. Now we're getting somewhere. This Farouk will be questioned much more thoroughly this time."

He sidled up to the cook and tapped him on the shoulder.

Chapter Twenty-two

Same day, late morning, Luxor

Zara Farouk had her own technique for getting past the guard at the front of the house. She put on her nicest smile, and as she handed the man (who was distantly related to Uncle Mustafa) his baksheesh, she explained that she just had to pick up a few necessities for the household. He did not even ask to see her vaccination site.

Zara ran through the streets of Luxor with her light canvas bag bouncing on her shoulder. Inside were clean underwear and a few toiletries. She had no intention of betraying her escape plans by carrying too much.

Zara's mother Mina had been unbelievably helpful. She gave Farzad money to have his motorbike serviced and helped him siphon off enough gasoline from Zawi's car to get them to their destination. She provided Zara with a cover story for her rendezvous with Farzad: the morning shopping, picking up ingredients for a vegetable stew. That's what she would tell anyone who asked her. Actually, Zara was on her way to Farzad's alabaster shop. Instead of eloping at night, which was what everyone would expect, they planned to leave in broad daylight, when the streets were crowded with people doing their daily errands and men drinking coffee and reading thc newspapers.

Her heart sped up from a heady mix of exertion and excitement. By the time her family sat down to lunch at home, she and her lover would be on the road to Aswan. No one would miss her for hours, until she was supposed to cook the supper. She reached the intersection near the Hammadi shop and slowed to a walk. It wouldn't do to appear to be in too much of a hurry.

Zara saw Farzad's bike parked along the narrow alley in back of the shop. The saddlebags were stuffed; she hoped they contained some food for the journey and not just Farzad's precious CD player and discs.

Farzad stepped out of the shop's back door and saw her coming. His face lit up with a huge grin as he advanced to meet her. "You're here, right on time! Now, I go to fetch a couple of pieces of alabaster from the storeroom as my father asked me. We will leave as soon as I return. He's going to say I've gone to the bank."

Zara laughed. "And I am shopping!"

Farzad crossed the alley to his father's storage shed and unlocked the door. As he disappeared inside, Zara stood twisting her hands together, assailed by a sudden sense of urgency. She moved over to the shed and peered inside. "Farzad, do hurry! I'm getting scared. I shouldn't even be here. It is too close to my father's favorite café."

"Relax, I will only be two minutes," said Farzad, closing the shed and walking back into the shop.

Zara made room for her meager possessions in one of the motorbike's saddlebags and strapped it shut. Then she heard men's voices.

"I have some new stone that just came in, my friend. It is a honey-colored alabaster, streaked with white."

"Actually, I'd like to see that maroon-colored marble you showed me out in your shed last week. Let's go look at that again."

She knew that voice. Zara looked around frantically for a place to hide, but found herself hemmed in by storage sheds and rubbish bins. She darted down the alley.

"The shed's hopeless right now; I've got Farzad rearranging it. I have a piece just like the one you admired here in the office."

The beaded curtain at the rear of the shop parted and Zara's father came through, accompanied by a shorter man wearing an olive gelabiyah.

"Zara! What are you doing here?" yelled her father as he spotted her running away. "Come back immediately!"

She heard his footsteps pounding down the alley behind her and ran faster. He grabbed her and whirled her around. She met his outraged glare.

"Why are you here, in this part of town? I thought you were in the vegetable market, shopping for your mother!" said Zawi Farouk.

Behind him, Mr. Hammadi made despairing motions at Zara and mouthed the words, "I tried to stop him from coming out here."

Zawi's attention was caught by the motorbike with its bulging saddlebags. "Just what is going on here?" he said in a terrible voice.

Mr. Hammadi cleared his throat, but his son Farzad came out of the shed and forestalled him. "Zara and I are leaving today," he said defiantly. "I will not let her marry your elderly partner, someone she does not love."

"You what? How dare you!" Zawi grabbed Farzad and twisted one arm up behind his back. "You'll do nothing of the kind!"

"Get your hands off my son, you miserable bully!" yelled Mr. Hammadi.

Zawi Farouk was past listening to anyone. He kept hold of Farzad with one hand and slugged Mr. Hammadi in the face.

"Hey! What do you think you are doing?" cried another man who was Hammadi's business partner. He streaked across the alley and launched himself at Zawi Farouk.

Men came running from other shops and suddenly a full-fledged fight erupted.

Farzad was down, buried under a pile of men cursing and pummeling each other.

Zara stood aghast, trying to think of something she could do to break it up. She grabbed a bucket of garbage from someone's back door and heaved it over the fighters. Two of them lifted their heads, sputtering and scraping at the junk on their faces, but then they dove right back into the pile of cursing, flailing men.

By the time the police arrived, the worst was over, and disheveled men were beginning to pick themselves up and assess the damage. There were several bloody noses and a few sprains among the neighbors who had joined the fight.

At the bottom of the pile was Farzad Hammadi, white-faced and clutching his right arm. "I think it's broken," he said, gasping as the pain hit him.

"My God, he's bleeding!" cried his father, who had escaped with no more than a black eye and a ripped shirt. He pointed to his son's left side, where blood was seeping out onto the dirt.

Another man ripped open the shirt to examine the wound. "He's been stabbed!"

Everyone gathered around again so Zara couldn't get through. "Let me in!" she sobbed. "He is my betrothed!" Her breath came in shallow gasps, and her heartbeat thundered in her chest.

"Okay, back away," said a new voice. "Let the medics through with a stretcher."

Mr. Hammadi put an arm around Zara as Farzad was gently lifted by four men, laid on a stretcher, and carried to the end of the alley where the white ambulance waited. "Don't worry, dear, we will get help." Then he looked around. "Where is that scoundrel, Zawi

Farouk? He started this mess."

"Gone, of course." said a neighbor cynically. "I haven't seen him for at least ten minutes."

"He will pay for this. I will make a full report to the police and then to the hospital," said Mr. Hammadi with grim relish. "Zara, my dear, you must go home. I will call you when I've checked on Farzad."

Zara slunk away down the alley. Where was her father? Would he go back to the house? She cringed at the idea of meeting him again. On the other hand, maybe her mother had an idea where she could hide until Farzad was better.

She quickened her pace, her sandals with the gaudy fake jewels slapping the pavement.

Chapter Twenty-three

Wednesday morning, September 15, Luxor and Karnak

A pale orange sun hovered on the horizon, ready to lift its bright head and heat up the desert and everything in it to cooking temperature.

Lisa abandoned the twisted sheets of her hotel bed after a restless night filled with dreams of Arabs on horseback chasing her, and James on his deathbed. After a shower, a quick cup of black coffee, and a few bites of a stale almond pastry, she took a taxi to Karnak. Perhaps she would have a little quiet interlude before the tourists swarmed like locusts all over the site.

The CDC doctor had told her she could telephone twice a day and that they would be able to assess James' condition better after his fever broke. Greg had promised to keep her busy so she wouldn't worry herself silly. Seemed he thought that sending her on wild goose chases would help. He was wrong: nothing would keep her mind from chasing its tail with lurid pictures of James dying—or living, but sightless and disfigured with terrible scars.

Mark Leonard, the CDC contact, tried to help, too. He assured Lisa—for the second time—that James had some protection from the vaccination, but Lisa found this assurance hollow. Even if James did not die like Safaa Farouk, there was quite a nasty spectrum between mildly ill to nearly dead.

Lisa was a champion worrier. Give her a situation, and her mind would immediately imagine the worst outcome. This time, she had little to go on and no one to talk to except the two men who were trying to cushion her fears and play down the danger. She settled on a feeble mantra that she whispered to herself as she strolled: "he will be okay, he will be okay..."

The early morning air was still cool and fresh, a welcome change from the stifling heat of afternoon that was the usual lot of Westerners herded like sheep around the archaeological precinct. Tourists, like "mad dogs and Englishmen," traipsed over uneven ground in the hot sun nearly every day while the natives slept.

Lisa paid for her ticket at the window near the entrance to the Karnak temple complex.

People on package tours tended to meekly accept impossible schedules, crammed with too many attractions and not enough bathroom breaks. After a dozen temples and an equal number of eastern markets, no one could remember the details of what he'd seen. Even for Lisa, who'd read extensively before her trip, keeping track of the various sites, obelisks, frescoes, and temples to various gods was a challenge. Lots of photos accompanied by scribblings in her spiral-bound green notebook would help her—providing, of course, that she could decipher her own handwriting. One of the problems of the computer age was that everyone developed calluses on their fingertips from tapping on keyboards while losing the muscles that made good handwriting possible.

Like a white rat running on a wheel in its cage, Lisa's tired brain returned again and again to smallpox and James. What were his chances, really? She was afraid to ask. When she returned from Karnak, she'd call again and find out how he was doing.

Maybe if she concentrated on the archaeology in this fabulous place, she could buy a few minutes of distraction. She pulled out the site map from her bag and spread it out on a handy chunk of sandstone. She'd passed the ram-headed sphinxes on her way into the site, intending to return after she had walked around the Sacred Lake again and taken some more pictures of the Hypostyle Hall. Oriented, she folded the map more or less neatly and extracted her water bottle. Although the temperature was still far from hot, her dry throat bade her to drink, drink, drink before the dry heat sucked out her vital juices.

She'd go to the back of the Amun complex first, the third or fourth court from here in front of the Great Festival Hall of Thutmosis III. Lisa crossed into the forest of huge, papyrus-shaped columns that made up the Hypostyle Hall, ignoring the sleepy guard who beckoned her to come further in. No doubt he would show her a perfect photo op and then expect payment. As on her first visit, she thrilled at the monumental size of the pillars; surely there was nowhere else in the world with such fat, beautifully decorated columns. And it was much better to wander through them with hardly anyone else around.

Lisa walked slowly, her beige Rockports making no sound on the sand-covered floors. She passed a set of tourist bathrooms. Unlike other sites where the preferred structure was a white trailer strategically positioned near a souk, this bathroom complex had been built of tan concrete so it blended in with the ruins.

Near the Fourth Pylon, Lisa could see her goal ahead in the dim light.

In front of her was the tallest obelisk in all Egypt, the one raised by Hatshepsut in honor of the god Amun. She thought about the sheer amount of labor represented by the granite and sandstone carvings all around her. An army of workmen, just like the folks who built the pyramids at Giza. They'd used ramps and ropes, not bulldozers and cranes, and still managed to build structures that wowed the world thousands of years later.

Next to the obelisk, at the angle between two towering walls, she could just make out the golden arches of a MacDonald's in the distance. The perfect juxtaposition of old and new.

A flicker of movement caught her eye off to the right. It was barely eight o'clock, so no tourists were here yet. So who was it, flitting among the obelisks at this hour?

Abandoning her earlier plan, Lisa turned towards the Sacred Lake.

A break in the temple enclosure enabled her to pop out near the second obelisk of Hatshepsut, now stretched forlornly on the ground. She looked around her, but no human being disturbed the morning quiet. Whoever it had been, that person had probably vanished into another entrance or around the corner of the enclosure. Now what?

Lisa smiled at the sight of the giant scarab beetle that she and James had circled only days before. That reminded her of James, and she replayed their pre-breakfast phone conversation as she walked around the lake.

"How are you, really?"

"I don't feel as bad as I did. My fever is down. I still have this damn rash on my chest and arms. And there's just the one sore on my forehead. But I can walk, talk, even eat a little."

"Only one pustule? That's better than some of the other patients, isn't it?"

"Sure. But each case is so different; they won't know for days what's really going on with me."

Lisa felt better after talking to him, but she knew that smallpox could go from "not bad" to very bad indeed in a matter of days. "When will they let me see you?"

"Well, maybe we can stage a meeting at twenty paces from the clinic, or whatever the other docs think is safe."

She didn't really expect it to happen. Most likely, James would be kept in isolation until the disease had run its course. She and James had a brief chat about extending their visas, which Lisa promised to take care of. Greg owed her one; he could help her find the appropriate official in Luxor or facilitate the process in Cairo.

Now, three hours later, Lisa had a ghoulish desire to see the site where Greg's agents had found Ali Farouk's body on the south shore of the Lake. No doubt the team had already removed every scrap of evidence.

Lisa found nothing on her circuit, but she had hardly expected to succeed on the first try. Consulting her guidebook again, she saw that she was very close to the Temple of Mut, the site of a series of granite statues honoring the lioness goddess Sekhmet. Always fond of cats, she enjoyed seeing the furry creatures venerated in stone. Cats had reigned supreme in ancient Egypt, and their little bodies were mummified along with a host of other animals considered sacred manifestations of the many gods. All the felines Lisa had met in her lifetime acted like they were still kings and queens—tell the human slaves to pass the tuna, no excuses accepted.

She was admiring the sleek lines of a black granite cat when she heard voices. Glancing up, Lisa saw a small cluster of men near a cement structure that looked like a storeroom with several entrances.

Idly, Lisa observed them until the tallest man turned her way. She froze.

It was their security guard, Zawi. What was he doing here? No longer in charge of their group because it had been quarantined on the cruise boat, he'd presumably been reassigned elsewhere. But he was a Farouk, part of the same Farouk family that now had several cases of smallpox. Greg had never said he was investigating Zawi in particular, just that he and his agents were keeping an eye on the whole family.

And where were Greg's men? They were supposed to be keeping an eye on each tourist site. Greg had said he didn't have enough men; either he thought it was too early in the morning or his agents were deployed elsewhere.

Lisa glanced at the men with Zawi. Surely the shorter one was also familiar?

Just as Lisa realized her position was exposed, Zawi Farouk spotted her. With startling speed, he crossed the distance between them and growled, "What are you doing here? You tourists should all be on the boat."

Lisa didn't like the hard set of his face or the glitter in his eye. "I am exempted from quarantine because my husband and I were vaccinated right after the first case of smallpox. And that first smallpox case was a relative of yours, wasn't it?" She hadn't planned on saying that, but his reaction scared her.

Zawi gripped her arm just above the elbow. "So what? No one is exempt from this disease."

"Let go of me," Lisa said. She hoped he couldn't feel her pulse racing.

The security guard dropped her arm and adjusted his facial expression to the calm, inscrutable look she was used to. "I am sorry," he said smoothly. "You were—how do you say? Out of context."

And you were out of order, she thought. Still uneasy, Lisa rubbed her arm. "I came early to take photos. And I am an archaeologist, so you must understand that this site is of special interest to me." She didn't owe him any explanation, but she didn't want to be hustled away from the site. "Why are you here? Surely it's too early to provide security for tourists?"

"Some of my cousins work here, and we are visiting." Zawi Farouk smiled and bowed slightly, returning to his meeting.

That it was a meeting Lisa was certain. Her hands tingled—the equivalent of thumbs prickling, she supposed. Zawi's calm admission of meeting relatives was probably true, she thought, but that didn't mean their meeting was innocent. She moved away from the little group, reluctant to use the bathroom when they were standing right in front of it. Instead, she walked over toward an ancient, disused refreshment stand. Lisa gazed longingly at the torn picture of drinks and snacks on the closed kiosk and imagined drinking an icy cold Coke.

Suppose the Farouks were indeed connected with the terrorists as Greg suspected but couldn't prove? They could be doing their own scouting around, looking for ways to infect tourists with smallpox. They could fan out to the different archaeological sites, observe how many people used the tomb tickets—or even the general admission tickets, since in some places you had to show them more than once. Or what about the site brochures and maps? They, too, were printed on cheap, porous paper, and were opened and handled constantly by visitors from all over the world.

She thought of her imaginary soda and shuddered. One could even adhere weaponized smallpox powder to the paper labels on Coke and water bottles.

Passing behind a convenient set of pillars, Lisa looked back at Zawi and his colleagues. A fourth man had arrived with boxes on a dolly.

From where she stood, she could see them unpacking brown paper-wrapped bundles.

Then, as the first partially unwrapped bundle was set on the ground, she gasped.

Toilet paper. Good old TP.

Toilet paper—why hadn't she thought of that before? The perfect medium, used by every woman and many of the men who visited the site of Karnak—especially if they'd sampled dubious bazaar food during the previous twenty-four hours. Toilet paper that you had to take from the attendant, and pay for unless you were a brazen, hardened type like Lisa. Never mind that some of the attendants would become infected too. They were the lowest of the low, uneducated Bedouins who could find no other job.

Zawi Farouk unlocked one of the storerooms and the men carried the packages inside. Lisa edged closer, pulling out her battery-operated digital camera. Her hands began to sweat as she fumbled with the case.

Lisa stepped behind a statue and peered cautiously into the open room.

One of the men had opened a roll of TP and was sprinkling something on the unrolled paper.

She raised her camera and clicked the shutter. As she moved back behind the sheltering pharaoh, the edge of the camera clipped the stone.

Zawi Farouk's head snapped around at the sound. In a few strides, he was beside her again. This time, there was no mistaking the rage in his eyes. "Interfering bitch! Daughter of the Unbelievers!" he snarled. He snatched the camera from her hand and hurled it away. "Mustafa!" he hissed at one of his companions. Then he rattled off a stream of Arabic she didn't understand.

The men surrounded her. Up close, Lisa could see that Mustafa was the cook from the hotel boat, released by Greg because he hadn't found any proof of smallpox virus stored in the freezer. Another man stuffed a filthy rag in her mouth, and hustled Lisa around to the rear of the storerooms. Zawi unlocked another door and shoved her inside. One man produced some rope and they bound her hands behind her and then her feet.

Zawi put his face up to Lisa's, so close that the spittle from his lips hit her gag.

"Later, bitch. Later, I will deal with you."

Zawi slammed the door shut, leaving a metallic clang that echoed through the tiny room. Lisa lay slumped against the wall where she had fallen when Zawi shoved her inside. She eased herself upright, favoring the right side where the ropes were cruelly tight, and looked around the storeroom to see who or what would be her companions for the next few hours.

Toilet paper, mountains of toilet paper.

Chapter Twenty-four
Same morning, Luxor

Shortly before nine the same morning, Abdul slid out of the hospital bed and stood up. Immediately, he was dizzy from the fever and days of bed rest. He grabbed the bed rail and swayed as he struggled for balance. His head and chest itched horribly.

Abdul looked down at his body. He was clad in a light blue hospital gown, open at the back. Where had they hidden his clothes? Furtively he groped around, finally locating his gelabiyah wadded up in a drawer. Abdul listened. It sounded like an early staff meeting going on in the main part of the field clinic. He slipped out of the hospital gown but the motion made him light-headed again, so he sat down on the bed.

When the room stopped spinning, he pulled the robe over his head. No shoes. Well, he'd just have to do without them. His feet were tough, tougher than the rest of him in his diseased condition.

Abdul removed the tiny mosquito net—shaped like his sister's shower cap—from the top of his water pitcher and took a long drink. There was a mirror in the bathroom. He had only a few pustules. That wasn't so bad—he'd seen other patients with multiple weeping sores. He hoped the disease would not leave scars; the girl he had his eye on would not like them.

Just this morning, very early, a team had been to visit him. The doctor said he'd been lucky; his case was relatively mild. He'd have to stay another week or ten days to be cured, but Abdul wasn't about to obey that rule. He was too worried. He hadn't heard anything from Zara or Farzad—what was going on with them? His father, he knew, was planning an early wedding for his daughter and his business partner, and it was unlikely that he'd wait until Abdul was well enough to attend. And Farzad, the disappointed suitor and Abdul's best friend—what of him? With all his plans in disarray, what would he do?

The long drink had steadied him. Abdul stood again and to his great relief, felt no dizziness. He crept out into the hallway.

So far, so good. The murmur of voices from the tiny conference room continued, and no one stopped him as he stumbled out the nearest exit. Abdul was amazed that there were no guards on the place, but then most of the patients who'd come in the last couple of days were far too sick to think of walking away. At least two of the unlucky victims were Farouk cousins.

A cluster of bicycles under a nearby tree drew his attention. As usual, only two or three were locked down—the rest were giveaways, bikes on their last wheels that the youths of Luxor borrowed freely and rode everywhere. It was understood that you could always find another. Only Westerners were concerned with locks and combinations.

Abdul grabbed a bike and pedaled slowly, stopping frequently to catch his breath and wipe off his streaming forehead. It was a bad sickness, this one that could produce so much weakness in only a few days.

As he approached the Avenue of the Sphinxes, a wailing ahead of him caused a ripple of fear. It sounded like—it was—centered near the Farouk household. Abdul tried to increase his speed, but he was too weak. He hopped off the bike and shuffled along, gasping, until someone spotted him.

"It's Abdul! We must tell him!"

"The man of the hour!"

"What will he do now?"

This last query came from a neighbor, Mrs. Haddid, the one who favored purple head scarves. What was she talking about? He entered his house and found his mother in the kitchen.

Mina Farouk sat stony-faced, her eyes fixed on nothing while two of her granddaughters huddled near her, wailing. To her son's greeting, she said only, "You're back, but it's too late."

"Mama!" He grabbed her by the shoulders. "This is the first you've seen me in days since I took sick. I just returned from the House of Pestilence. What is wrong with you?"

Her brown eyes turned slowly onto his face, but the pustules did not move her. "Your friend, Farzad, is in the hospital."

Abdul dropped his hands. "What? How can that be? He was fine when I saw him!"

"And your sister has been forced into marriage with that baboon, that pile of camel shit. It is wrong, she does not even like him."

"No, I do not believe you! What happened? You must tell me."

Tears flowing now, her eyes alive with hatred, his mother told him. Farzad and Zara, knowing that her father would never approve the marriage, had arranged to travel to Aswan. There they planned to stay with a second cousin of Mina Farouk until Farzad found work. Zara had been packing their belongings into the saddlebags of Farzad's pride and joy, his ancient motorcycle, when Zawi had spotted her in the alley.

Then she described the fight and its finale, with Farzad being carried away by ambulance with a broken arm and a stab wound in his lower abdomen.

"How is he?" gasped Abdul.

"I think, all right. He has lost some blood, but he will mend."

"And what happened to Zara?"

"She returned to the house, and I was going to hide her with friends in another part of town. But your father came back too soon. He dragged her away, saying he would deliver Zara to Achmed's house, where they would be married by an imam within the hour. He and Achmed had already signed an agreement. There was nothing I could do."

Abdul's blood boiled as he paced the narrow kitchen. His already weakened body rebelled and he shook from head to foot. He ached for his sister who was now in a virtual prison, a marriage she didn't want. And Farzad! All his dreams in ruins. What a horrible mess.

"My son, Zara said...she said...she left you a note. She wrote it before all this happened," came the shaky voice of his mother. Her hands gripped him gently and he stopped pacing. He took the scrap of paper and read the hurried writing.

"My brother, we are eloping today. I cannot marry one I don't love. You, who love Farzad too, will understand. But you may be angry with him for taking me away, and for ending our plan to operate the shop together. Do not be; it is for the best. One day you can come and join us, and we three can share a home. Think of it: I will make your favorite dish of lamb with apricots, just for you, with baklava to follow. I will find a café and send email to the usual address when we are settled. Your sister, Zara."

Abdul smiled through his angry tears as he read the bit about his sister cooking for him. Zara had wanted for so long to have her own kitchen. She loved to cook special dishes. But his sister was by now the wife of a man old enough to be her father. She would be too distressed to think about cooking. And Achmed had other wives.

Out loud, he said, "First Aunt Safaa. Now Zara. How many more people will my father harm in the name of Jihad?"

Mina shrieked, "Safaa? What do you mean?"

"Sit down, mother."

Abdul collapsed in the other chair and told her everything he'd learned. From breaking the ampule on his aunt's scarf to Zara's account of the midnight meeting she'd overheard. How guilty he felt about his own role in aunt Safaa's death...

"You are not to blame!" said his mother fiercely. "It is the fault of the Farouk brothers to meddle in such things in the first place. Killing tourists with powder—it is crazy, it will bring us all down!"

She leapt to her feet and circled the kitchen, telling Abdul things he'd suspected but never fully understood before. How Zawi, Ali, and Mustafa Farouk smuggled antiquities and drugs in the past. They were partners in every endeavor. Uncle Ali thought he was in charge, but Zawi was the real ringleader. About Zawi's insatiable desire for money and the things money could buy—a new silk robe, a satellite dish, special tobacco for Uncle Ali. He had no honor, no integrity. Zawi was not fit to be her husband; she would leave him. And he would never let Abdul become a doctor...

She spun around. "Abdul, you must be the man of the house now. You must stop your father from doing this terrible thing."

Abdul was stunned. His mother was telling him to flout tradition, to dishonor his father? It was what he wanted to do anyway, but he'd never imagined she would give him her blessing. But how?

His mind blazed as the little seed planted by the Israeli agent found fertile ground. Of course! He would tell Greg Manzur everything he knew. Manzur would reward him and then Abdul could support his mother and still afford the university! His resolve grew from a sprout to a thick vine.

Abdul stood and loomed over his mother. The effect was spoiled slightly by his inability to keep his balance and he grabbed the back of his chair.

"What will you do?" she asked.

"I am going to talk to the Americans. They will take care of Papa and his associates. And then we will not have to worry, ever again."

Her eyes widened and she clutched his arm. "No! They will lock you up—or kill you!"

Abdul's voice was steady and loud. "No, Mama. I will not be killed. I have information to sell that three governments want."

"Arrgh! You sound like Zawi! How can you be sure?"

"This is not one of Ali's crazy schemes. This is different," he said. "My information will help the police catch them before they do their dirty work. It will be okay, Mama. Now let me go."

He chose not to tell her about the man who worked for Mossad—she would never believe anyone could trust an agent of the Israeli government. Nor did he tell her that he did not yet know exactly what his father was planning.

Mina Farouk eyed her son skeptically. "Well, you must eat something before you go."

Chapter Twenty-five
That afternoon, Karnak

What time was it? Lisa couldn't see her wristwatch because her hands were tied behind her. And she had no idea how long she'd lain against the concrete wall of her little cell, surrounded by brown-papered bundles. A narrow window above the door showed bright daylight still, so it was probably afternoon. How long was she going to be locked up with all these stacks of toilet paper?

Her arms ached from the ropes, and there was a cramp in her left calf she couldn't release. That was bad enough. What threatened to drive her mad was her parched mouth, blocked by the rough cloth Zawi had jammed in her mouth and then covered with another strip of cloth tied behind her head. Maybe, if she blew hard enough, the wad inside her mouth would dislodge and she could scream for help. Mustering all the breath she had, Lisa puffed out her cheeks and blew like a nursery picture of the North Wind. The result was a sound like a strangled goat. Her thirst increased.

Lisa pulled her knees in towards her and then stretched out her legs, hoping to ease the cramp. She did it again and the pain subsided.

Where were all the tourists? She was counting on traffic at the site to bring people close to her position in the storeroom. Then she could thump with her feet on the floor to attract attention.

But this storage complex was at the back of the site, far less visited than the restrooms near the ticket booth and the souk. That was where the tourists would linger, not here.

Her stomach rumbled loudly. She'd eaten nothing but part of a stale pastry and some coffee with milk that day.

Poor James, he was getting so restless. Now they were both in quarantine of sorts—but James was the lucky one. At least he had some human contact with the other doctors who were keeping an eye on him. Her mind slogged through the quicksand of worry again. How sick was he? Did he have a full-blown case of smallpox, with pustules erupting hourly? Would the vaccine really protect his life? What if…

Lisa shook herself. She wasn't helping her situation by brooding. She needed to get out of here. The terrorists hadn't killed her outright, but she had no idea what their plans were. Someone—probably the sinister Zawi Farouk—would be back to either silence her for good or move her to a more remote location where she had less chance of being rescued.

The room had a second door off to the left, suggesting an inner storeroom. But the closest door, the one with the little window above it, was a few yards away. With both hands and feet bound, she could only move on her butt in a sort of scooting motion. She raised her feet about six inches, shoved them forwards, and scooted her bottom up to meet them. The floor, like the walls, was made of rough concrete. Lisa wondered grimly how long it would take for her thin pants to wear through. The scooting motion wasn't working. She tried moving one hip at a time and using her heels for leverage. The butt-walking worked better, and soon she lay on her back, gasping for breath, with her bound feet resting on the door.

She rested for a few moments, trying in vain to ignore her thirst, and listened for any movement outside. Faint voices reached her ears and a jolt of adrenaline coursed through her body. Lisa began counting to herself and banging the door with her bound feet at irregular intervals. She moaned in between hits.

The voices grew fainter. Damn! Try again.

One, two, three…hit. One, two, three, four…hit.

Abdul's bike had no brakes and uncertain steering. Just like me, he thought—I'm so weak from the sickness that I can barely peddle in a straight line. He reached the ticket booth at the entrance of the Luxor Temple and parked the bike up against the concrete wall.

"Hey, Abdul! I thought the Americans had you in the House of Pestilence," said his second cousin Akbar.

"They did. I left," said Abdul, wiping his sweaty face on one sleeve. "Have you seen my father?"

"He was here about fifteen minutes ago. He and your uncle went towards the storeroom."

Abdul walked a little way until exhaustion forced him to lean against a handy sandstone Pharaoh. Although his body was weak, his mind was busy connecting the dots of all he'd seen and heard.

Manzur, the Israeli agent, said innocent people were dying. He also mentioned small packages, and a powder that made people sick.

He'd heard his father and uncles muttering about another "shipment," and from Zara he knew about the late night meeting discussing what had gone wrong the first time.

It sounded like they were waiting for more powder, probably in glass ampules like the one he'd found in Uncle Ali's tobacco jar. They were planning to put it on something that tourists handled.

So his plan was to steal the shipment and thus thwart his father's plan. But first he had to find it.

Abdul started walking again until his breath came in hoarse gasps. He wanted nothing more than to lie down in that nice, artificially cooled clinic and go back to sleep. Sleep for days, until the sores disappeared and he was well.

He paused near a rack of colorful brochures and site maps. Where was everyone? Usually, he'd see one or more of his cousins sweeping around the cold drinks kiosk or restocking the restrooms with paper products. It was still only the middle of the morning, so very few tourist buses had arrived.

Abdul traversed the almost deserted site, aiming for the storerooms at the back. As he reached the first locked room, he heard his father's voice inside. It sounded like he was on his cell phone.

"El-Hol here."

Abdul knew this was the second part of their pass-phrase in Arabic, the first part being "what is the name of the Father?" The complete phrase was the local name for the Great Sphinx of Giza, "Abu el-Hol," or "Father of Terror." How appropriate for Zawi to assume that title.

"Eh? You can move the stuff to the storeroom at Karnak. Yeah, the one near the restrooms. Mustafa has the key."

Abdul nodded to himself. Mustafa wasn't the only one to have a key—Abdul had one, too. He'd pinched it from Ali's pocket about a year ago.

He leaned against the door, careful to make no sound. As he listened, he smiled grimly. Now he knew exactly what to do to stop his father.

Greg was frustrated. Locating the terrorists was proving just as difficult as containing the smallpox outbreak. Like Dr. Mallory, he had to deal with agents and officials from several countries. They all had different ideas about what to do and how to do it. Greg was still "in charge," but the operation was like wrestling an octopus. Too many people involved, too many people wanting continuous updates. Cell phone operation was so haphazard that he'd been reduced to sending couriers. This exacerbated his other problem: manpower. Greg's Mossad agents were now deployed in Jordan, Lebanon, Israel and Egypt along with the agents from other governments. No one had enough people.

Rafi and another colleague had spent hours patrolling the key archaeological sites around Luxor at times when tourists were present. They returned at odd hours, looking for unusual behavior among ticket takers and restroom attendants, but there were only two of them. They simply couldn't cover all the territory, and this made Greg very nervous.

He lit a cigarette and inhaled deeply. The smoke calmed him as he returned to his car near the Temple of Luxor. Okay, time to sum up. What did they know?

That a local group was in cahoots with Hamas in Israel and Hezbollah in Lebanon. And, that they were using code names—various animals in Arabic—and something else no one had translated yet. So far the names of those involved had eluded detection.

The Farouk family had come under suspicion from the very first case of smallpox because of their reputation for petty smuggling. Greg had argued vehemently for bringing in one of the brothers and making him talk, but he was over-ruled by the CIA and Egyptian authorities. We need to track them, was the counterargument, and see where they go and whom they contact. We must bring down the whole organization, not just one or two individuals. Reluctantly, Greg agreed to this policy.

The cook on the Shehrayar, Mustafa Farouk, had been a case in point—his behavior, as witnessed by Lisa, had been suspicious, but Greg still had no proof he was involved in the smallpox plot. Since no package had been found in the freezer, Greg had been forced to let the man go.

Hamas, Hezbollah, and similar terrorist organizations, were like many-headed Hydras—cut off one head and another would appear. What scared him was that the whole thing smacked of an internationally organized, al-Qaeda kind of plan. Even if the CIA immobilized the local operators, their colleagues in other parts of the Middle East would proceed—in fact had already done so—with targets in other countries. And that meant Western tourists in Cairo, Amman, Beirut—hell, anywhere that a small container of smallpox dust could be transported—were being infected. The next obvious step was disease transmission by airline to anywhere in the world.

His cell phone rang.

"Have you found the smallpox dust yet?" he asked Rafi.

"No. Yosef, Itzak, and I have been casing every ticket booth around the entire Luxor area—the Temple, Karnak, the Valleys of the Kings and Queens, and the Hatshepsut site—but you know there aren't enough of us to be at any one site more than a few hours a day. But boss, we have a lead on one of the Farouk cousins who may know something."

Greg gripped the phone tightly as Rafi described the set-up—a secluded house at the end of a dead-end street near the main souk. He made up his mind to ignore the instructions to watch and wait.

"Let's go get him," he said. Act first, explain later. It was a philosophy that had worked many times before.

"How many men do you want?"

Greg thought rapidly. "Four. Can you manage that?"

"Okay, but I'll have to pull Itzak away from Karnak. Oh, and Boss? I phoned Lisa Donahue as you asked me to. She isn't there. In fact, no one's seen her since breakfast."

"Damn," said Greg. He slid the phone back in its holster and ran for his car.

Chapter Twenty-six

Same day, afternoon

"He who does evil, the (very) river-back abandons him, and the floodwaters carry him off."
(The Instruction of Amen-em-opet)

Zara Farouk sat dejected at the kitchen table in Achmed's house. A bowl of lentil soup stood before her, but it was untouched. Her face felt hot and swollen from crying, and she wished she were dead.

The past two hours had been pure nightmare. She and her husband-to-be were awaiting the arrival of the imam who would marry them. She would soon be the wife of an older man she despised while the man she wanted was bleeding and hospitalized.

"You must eat something."

She looked up at the bearded man across from her. He was old and overweight but not as ugly as she had first thought. His eyes were kind and his voice surprisingly gentle.

"I can't," she gulped.

"Tell me what is wrong. You have said nothing since your father left. I hate to see you so distressed."

Zara stared. He sounded as if he really cared how she felt!

"My father is a crook and a cheat—and he tried to kill Farzad."

"This is not good. Who is Farzad?"

She took a deep breath. "Farzad Hammadi is my true love. The man I really wish to marry."

Achmed frowned. "Then your father has been playing with me. He implied that you were willing. And I did not know he tried to kill your…intended. What happened?"

Gradually the story came out. When Zara arrived at the scene behind the alabaster shop, her hands crept up to fiddle with her headscarf and her eyes filled again.

Again, the man surprised her. Achmed rose from his chair and handed her a large, immaculate handkerchief.

"Do not worry any more," he said. "I am going to help you."

Zawi Farouk unlocked the door to the largest storeroom at Karnak, one that was several rooms away from where the American woman was imprisoned. Moving confidently, he crossed to the box in the far left corner where he had left the shipment of ampules tidily hidden under burlap sacks. He halted as he saw two of the sacks on the floor. Snatching off the remaining cover, he cursed as he realized that the package was gone.

Who could have done it? Everyone who had a key worked for him now. Had Mustafa been before him? He heard a sound behind him and whirled around.

"Mustafa! Do you have the powders?"

Mustafa's eyes grew huge. "No, isn't the package there, right behind you?"

"No," snarled Zawi. "Allah help us if we don't find it quickly."

Mustafa averted his eyes and shuffled his feet.

"What—did you see something?"

"Your son, Abdul—Khalid told me he was here. About an hour ago."

Zawi felt the rage building. It could not be Abdul! Yet even as the denial crossed his mind, a little voice told him it could indeed be his alienated son. Abdul, who so foolishly took the side of the women in most things. Abdul, who scorned the family business that was their livelihood.

He gave Mustafa some instructions on how to gather the other men. Then he ran for the truck, cursing again as he thought of the disastrous timing.

Twenty minutes later, Zawi killed the engine in front of his house and slid out of the driver's side. Looking around, he saw a bicycle propped against the side of the house and his pulse accelerated. If that young fool were here, in the house, he'd lose no time in forcing a confession and recovering the package.

Abdul sneaked into the Farouk house on the Avenue of the Sphinxes, intent on hiding the package before anyone caught him and maybe grabbing some lunch. He stood still and listened carefully. Everyone was out, and all he could hear was the ticking of the cheap, wind-up clock on top of the miniature fridge. He slid through the bead curtain into the kitchen and extracted the package from his robe.

It was larger this time, brown paper sheltering a bubble-wrapped inner bag. Gingerly he unwrapped a few layers, praying that the handling of it could no longer make him sick. He let out his breath as he saw them: about fifty small, amber colored glass ampules. He smiled as he slid the resealed package into Uncle Ali's four-foot tall tobacco jar. Surely his father would never think of this hiding place! He would look all over Karnak and the Temple sites, never suspecting his son would dare use the same place as his uncle.

Abdul opened the fridge. His appetite was still far from normal, so he took a small piece of pita and scooped up hummus with it. While he ate, he sat down and planned his next move.

He'd use the borrowed bike again to return to Karnak. Then, as soon as he found uncle Mustafa and the others, he'd…

A small sound distracted him and he turned. His father Zawi stood just outside the beaded curtain. How long had he been in the house? Had he seen Abdul hide the package?

Abdul leapt to his feet, knocking over the chair. Zawi surged through the curtain, grabbing his son by the front of his gelabiyah.

"Where is it?" he hissed, drawing his knife from its inner pocket and bringing up to Abdul's throat in one smooth motion.

"I don't know what you're talking about, Father," gasped Abdul, shutting his eyes as the tip of the knife pricked his skin.

"Look at me!" screamed Zawi. "Mustafa saw you this morning! You have the package and we need it, right now!"

Abdul twisted in his grasp and his eyes slid inadvertently towards the tobacco jar.

Zawi smiled evilly and deliberately pushed on the knife so a thin trickle of blood appeared on the boy's neck. "Aha! So you thought you could trick your father…"

"Stop right there! Or I will kill you myself!"

Zawi turned, still holding his son at knifepoint. "Achmed! What are you doing here?"

His business partner, a bull-necked, pot-bellied man of middle years, oozed into the little room. In his hand he held small pistol. Behind him was Zara, gripping her scarf with both hands held under her chin.

"I am here to return your daughter. I take no wife who is not willing, and you never told me she loves another or that you almost killed her lover! The ceremony was not performed; I sent the imam away. I will also return the dowry, since I have no use for it—or for you as a partner. Now, let him go."

Zawi released his son and put the knife on the kitchen table. "Listen, Achmed, wait. I'm sure there is a misunderstanding. My friend, we have made so many plans together."

"Indeed." Achmed spat on the clean floor. "But I will not work with a suspected murderer, one who everyone is saying killed his own brother, and who threatens his own son with mutilation or death!"

Zara moved forward, her eyes enormous but very sure. "You have better friends than you deserve, Papa. He was very good to me while I was at his house. I told him everything."

"You!" screamed Zawi. "Are all my children so ungrateful? If you would just do what you are told!"

Achmed began to unwind his turban with one hand and motioned to Abdul to hold his father's hands still so they could be tied behind him. The gun remained pointing at Zawi's chest.

Zara took advantage of her father's temporary immobility. "I know now you did not kill Farzad," she said. "But you tried, and because of that I owe you nothing, nothing!" Her face changed and suddenly she was no longer a timid girl but a woman full of righteous wrath.

"My Farzad is out of hospital, and he is coming soon to take me away. We will move far away and I will sell alabaster while he finishes school during the day. Then at night school I will do college courses—yes, me!—and learn accounting so I can help him with the business!"

Achmed, holding the long strip of cloth from his turban, stepped closer and sneered at his former colleague. "You sorry son of a jackal," he said. "The only thing you can do right is produce beautiful daughters!"

Zawi's face contorted and he moved with the speed of a striking cobra. He punched the gun out of Achmed's hand, shoved his daughter out of the way, and leapt for the tobacco jar. Abdul grabbed his arm and was flung against the wall.

His father scooped up the package, scattering tobacco everywhere, and fled through the beaded curtain.

Abdul seized the knife and followed him. He careened around the house just in time to see Zawi, still clutching the package of smallpox ampules, zoom away in his truck.

Behind him, he heard Achmed yell for the police.

Chapter Twenty-seven

Same day, near sunset, Karnak

The sun was sliding down the sky like a ripe peach rolling across a table.

Abdul rode as fast as he could, sweating copiously as he wove away from his family's house on the Avenue of the Sphinxes. He still felt weak but his head was clear. Perhaps the urgency of stopping his father had given him the medicine he needed. There were narrow alleys only a bicycle could traverse, a shortcut to Karnak.

Just as he was approaching the site, the chain on the bicycle popped off and he skidded to a halt. Abandoning the bike, he shuffled past the ticket booth, towards the storage and utility area near the sound and light show seats, casting his gaze around for the cousin he knew was a guide. No doubt Khalid was mopping up with a few stray tourists, demanding money for suggesting the perfect photo op: sunset viewed through the famous pillars of Karnak.

Warily, he approached the group of storerooms, wondering where his father and his associates were. As he neared the room where the ampules had been hidden, a thumping distracted him. He listened, puzzled. It seemed to be coming from the smallest storeroom at the end of the row. He walked closer and heard a snuffling sound, then a moan. Someone was inside!

Abdul felt for the knife he'd taken from his father and ran to the guard's break room where he knew the keys were kept. No one was there, so he grabbed the keys and returned to the door. In seconds he had unlocked it and opened the door.

As the door opened, Lisa's feet crashed down onto the doorstep. She winced as her body was jarred from heel to chin. The setting sun was in her eyes, so she couldn't see the face of the man who'd unlocked her prison.

Then the gag was ripped out of her mouth, and she saw the

pox-marked face of a young Arab.

"English?" she gasped, barely able to make her dry mouth work.

"I speak some. Enough," answered the young man. "Hold soft so I cut rope."

He pulled out a murderous-looking dagger and proceeded to cut her bonds, helping her rub her wrists and ankles so she could stand up.

Lisa stood up with difficulty, feeling shaky. "Water?" she croaked, making drinking motions.

Her companion pulled out a battered tin flask. Lisa closed her eyes and took a long drink, surprised to find that the contents were actually water.

She opened her eyes again and noticed something familiar about him. "Who are you?"

"Abdul Farouk. Zawi Farouk is my father." He said this with a snarl and a curling of his lips that boded no good for the father.

Lisa wanted to believe he was a friend, but she had to be sure. "He locked me up. Do you know what he is doing?"

"Trying to make tourists sick. I know not how he is doing this. But he works with my uncles." His thin face hardened into a mask and he looked much older.

"Are you sick, too?" asked Lisa, focusing on his facial sores and remembering the discussion about a young member of the Farouk family being the second smallpox victim. She didn't know much about smallpox, but this guy looked like he still belonged in hospital. At least she was protected by the vaccine—she hoped.

"Yes, but I okay now. Doctors saved me. And an agent came to visit me in clinic. He asked many questions about sickness."

Could have been Greg—or someone else. Lisa had no way to find out. She had to make an ally out of this youth, no matter how sick he was. "Listen, Abdul. I think I know what your father and uncle are planning. Will you help me stop them?"

Abdul nodded. "Now come," he said urgently. "We go to another place to make plan. They must not find you here."

A short while later Lisa and Abdul sat in the sand, leaning their backs against an obelisk that was lying on its side. Dusk had fallen quickly but a dim glow from a streetlight off to the right made it possible for them to see each other. Abdul had told Lisa that there was no "Sound and Light" show scheduled tonight, so there would be no tourists and no guards.

Lisa was exhausted and filthy. She finished the lifesaving bar of chocolate Abdul had produced, like a conjurer, from a pocket in his robe, and took a deep breath. "Abdul, remember the toilet paper in the storeroom? It's really porous stuff. From what I saw this morning, I think the plan is to contaminate the paper, then put it in bathrooms to make tourists sick."

"Con-tam-in-ate?" He stared at her lips.

"Infect. With smallpox powder."

He stared at her.

"Smallpox is one of the most dangerous diseases in the world. It is a virus—a small particle—that has somehow been converted to a powder. Once someone inhales the powder into his lungs, he gets sick. Like you, I think."

Abdul clearly did not understand, so Lisa tried explaining again, using more words and speaking slowly.

Abdul's eyes grew enormous. "My aunt Safaa...Allah forgive me," he whispered, closing his eyes tightly. His face paled as she watched him, wondering exactly what he was remembering.

He opened his eyes again and looked at her soberly. "The powder comes in little glass...pitchers?" His hands made a shape in the air.

"Bottles? Flasks?" It didn't really matter; she understood that they were small glass containers.

"One broke when I fetched it out of Uncle Ali's tobacco jar," he said bleakly. "Powder fall on headscarf. And my aunt Safaa...she died."

Lisa clapped her hand over her mouth. The index case! Safaa Farouk was the first smallpox victim—and now she knew how it had happened.

"And I got sick. From being close to Aunt Safaa?"

"Probably. Or by inhaling some powder when you broke the glass container."

Suddenly they heard voices. Both of them peered around the pointed end of the obelisk.

A group of trucks and men were gathered near one of the refreshment kiosks. Abdul whispered that one of the men was his uncle Mustafa. Lisa maneuvered so she was on her stomach with just her head sticking out, next to Abdul's. They watched as the men transferred brown paper bundles of toilet paper from the storeroom next to her former prison to the trucks. Then Abdul gasped as he saw Zawi Farouk roar up in a Fiat and hand over another brown package.

Lisa pulled Abdul back so they were both completely hidden and looked at his horrified face. "What is it?"

"They have it now!" he groaned. "The stuff that makes people sick! There are little glass pitchers in that package, full of evil dust." Another hurried conversation ensued and Lisa realized the terrorists had not yet inserted the powder into all of the TP. They still had time.

Abdul grabbed her shirtsleeve. "We stop them," he said, eyes bright with purpose. He looked around and then said, "Wait here, nice Miss."

The young man sneaked around the obelisk, behind a row of statues, and over to a truck that stood all by itself. Lisa watched as he whipped out a tube of some kind, unscrewed the gas cap, and sucked out some gasoline into the water flask that he pulled out of his gelabiyah. It certainly looked like Abdul made a habit of stealing gasoline. Glancing her way, he lifted his hand and ran off to the left, where a small building loomed.

Lisa waited nervously, alternately watching the truck loading and keeping an eye out for Abdul's return. The men were nearly through; if she and Abdul were going to stop the dispersal, they would have to move fast.

Abdul reappeared with a canvas sack. Soon he was behind the obelisk, triumphantly unloading his loot. Two Coke bottles and a crummy old cotton rag. He pulled out the flask and unscrewed the top, pausing to catch her eye. "You see?" he said, with his gaze steady on her face.

Bottles, rag, gasoline. A way to stop trucks. She sucked in her breath. "Molotov cocktails? Gas bombs?"

"Molotov, yes! You smart lady!" His smile transformed his features, and Lisa forgot about the smallpox scars.

Lisa wondered what would happen to the smallpox dust if they succeeded in blowing up the trucks. Would it vaporize and become harmless, or would it escape burning, become airborne, and spread faster than if they did nothing? There was no one to consult and only one way to find out.

"Wait," she said, as Abdul raised the flask above one of the bottles. She pulled out one of the index cards she always carried in her pocket for notes and folded it into a funnel over the narrow mouth of the bottle. "Now pour."

Abdul neatly divided the gasoline between the two Coke bottles. He ripped the rag into two parts and made rough stoppers for the two bottles. Grunting when he realized he didn't have enough cloth, he whipped his kefiyyeh off his head and tore off a piece of that. With another devastating smile, he handed Lisa her bomb.

Chapter Twenty-eight

Same day, late evening, Karnak

Zawi Farouk wiped the sweat off his brow with the sleeve of his robe.

They had finished dividing the ampules and the toilet paper and loading two trucks, one for the Valley of the Kings and one for the Temple of Hatshepsut. To disguise the cargo, the workers had thrown colorful cotton shirts on top. These would be delivered to the market stalls while the TP went to the bathrooms. The attendants, poor dupes, would break the ampules with knife handles or Coke bottles and scatter the dust when they made up their neat little packets of TP to hand to each gullible tourist. He smiled as he thought how the tourists would pay for the privilege of dying horribly. There was enough dust to infect dozens of them, and they would carry the disease to other countries by airplane.

He glanced at his watch. Barely nine p.m. Now it was time to deal with that American bitch. He pulled Mustafa aside for a quick discussion.

"Where can we hide her until later? I don't want her killed outright. My plan for this Unbeliever is to release her in the desert near Kharga. If she makes it back before she dies of thirst, good for her. If she doesn't..."

Mustafa laughed. "Then our government has one more dead tourist to explain. I suggest the house of my friend in Al-Mahatta Street. He has a shed where we can hide her."

"Okay. You go get her. Take my car. The trucks must deliver their cargo before dawn."

The two men separated, and Zawi strolled over to give final instructions to his drivers who were lounging around the tourist café drinking Cokes and smoking Turkish cigarettes.

As he reached the café, he heard the tinkle of breaking glass. Whipping around, Zawi was in time to see one truck catch on fire with a whoosh of flames. A clink near the second truck and more flames erupted, this time on the ground near the front right wheel.

He screamed at his men, who jumped to their feet in confusion. Two started towards the vehicles, but it was clearly too late. The first truck, and its lethal contents, was blazing away merrily. The heat was unbearable.

Furious, Zawi jumped into the second truck and reversed it away from the first so it wouldn't catch on fire. He hollered at the driver to take it away now, and split the load between the two archaeological sites. Then he looked around for the cause of the fire.

To his right, on the roof of one of the storehouses, he spied two figures. One was the disheveled tourist, the woman he had locked up.

The other was his son, Abdul, who yelled at him as soon as he saw his father looking at him and raised his arms high. "That is for Zara and Farzad, you son of a jackal!"

Zawi felt a rush of blood to his face. "How dare you! You compromise the whole family! Our livelihood is at stake because of your foolishness!" He ran up to the building, glaring at his son above him, his fingers hunting for the knife that was usually in his pocket. It was gone!

"No amount of money is worth destroying my sister's life! You should never have tried to marry her to that filthy, aged business partner of yours! I will see justice done." Abdul was clearly prepared to continue, but the American woman grabbed his arm and pointed.

Zawi Farouk turned around and discovered that two Egyptian policemen, accompanied by a slight man with floppy, dark hair—must be the Mossad agent Mustafa had warned him about—were running towards him. Snarling, he glanced at the funeral pyre that was the first truck. Just then, the gas tank, finally heated up by the flames, exploded. Zawi looked for the second truck.

It was gone. Khalid had succeeded.

Zawi dashed for the back door of the Karnak complex that few people knew about.

Lisa Donahue stood under a warm trickle of water. Her eyes were closed and she could barely support herself. Too bad this hotel didn't have the sit-down showers so popular at Cairo hotels. She'd sit down and never get up until the hot water ran out.

The phone rang.

Grabbing a towel, she stepped out of the miniscule bathroom and picked it up.

"Lisa! How are you?"

She smiled. It was Greg, sounding uncharacteristically concerned.

"Not too bad. A little sore here and there from the rope burns. Thirsty—I've already drained two bottles of water. Some cold beer and a sandwich—or something more solid—would complete my recovery." She gripped the phone tighter. "Did you catch the driver?"

"No. But we found the second truck."

"Where?"

"At the temple of Hatshepsut. It was empty."

Lisa was motionless. "And the ampules of smallpox dust?"

Greg sighed gustily into the phone. "Gone. I'm afraid they succeeded with at least part of their plan."

"Not with toilet paper, surely?"

"Probably not since Hawass has closed all the sites to tourists for the next two weeks. But there are so many other ways they could distribute the stuff."

Lisa remembered all the ways they'd discussed—sprinkling dust over carpets, papyrus, paper tickets, labels on water bottles. The possibilities were endless. "I suppose you can't keep the tourists out of the souks?"

"Hardly. Especially now that the archaeological sites are off limits. But we've sent flyers and emails to every local tourism agency. I don't really believe they'll get the word out in time."

"I don't, either. What about the airlines?"

"You mean, to keep tourists from flying home while infected? We're working on it. I think the airlines will implement a shut-down in the next twenty-four hours, but a few flights will still slip through."

Lisa shivered as she thought of tourists returning to New York, Boston, London, Berlin, and Tokyo. Which ones would start an epidemic in their hometowns?

Greg interrupted her train of thought. "Why don't you come down to the bar and I'll buy you a midnight supper? We can talk some more then."

"Okay. But I need to phone James first. He didn't answer the first time I called, and I really need to hear his voice."

"James!"

"Thank God! Manzur left me a cryptic message that you were safe, which was worse than nothing because I didn't know you were in danger! I thought you were going to the Luxor Museum this morning. I've been tearing my hair out wondering what was going on." His voice cracked.

She'd lied to Greg. Her harrowing experience had caught up with her and she ached all over. She took a shaky breath. "I'll tell you everything, but first, how are you?"

"No change, really. A couple of pustules, one on my forehead and one on my chest, but the fever is down."

"I suppose you're going to tell me I can't relax yet because everyone gets sick a little differently?"

"That's right. We have to wait a few more days. Now tell me what happened this morning."

Lisa sighed with frustration—she'd really wanted to hear more positive news. "Well, after we talked, I changed my mind about the museum. I took a taxi over to Karnak."

James listened with only a few expletives until she got to the part about the Molotov cocktails. "Jesus! Do you have any idea how dangerous those things are? Quite a few people get burned when the flame shoots back into the bottle before it's thrown!"

"I knew it was dangerous, but at the time, it seemed the only way to stop the Farouks."

"Yeah. Guess you had to have been there. Still, you seem to be terrorist-prone, or maybe just unlucky. Or you're like that damn cat at home, with eight lives spent and one to go."

Lisa laughed. "You're comparing me to Oreo? Remember, I have a great advantage over him—my brain is several times larger than his."

James muttered something into his beard that didn't travel over the phone.

"Better I don't know what you just said. James, what will the CDC and WHO do now that some of the smallpox virus stores have disappeared again?"

"Wait and watch. They'll assume some virus has been released by the Farouks. Then, they'll isolate as many tourists as they can for as long as they can and look for symptoms. It will be ring containment all over again."

Lisa nodded to herself. James had explained the method used in the 1970s to isolate progressive circles of intimates and acquaintances to keep smallpox from spreading. The idea was to starve the virus; if it had no new human hosts, it would die out.

"I bet the local tour guides will have an awful time with all the cancellations. Our group gave Mahmoud hell at the slightest change in plans."

"Wait until the embassies have to deal with all the people trying to get back home and being told they have to stay in Egypt for two weeks.

I wouldn't want any of the PR jobs now."

"James, I'm starving and Greg promised me a cold beer and some food downstairs."

James snorted. "I suppose he's allowed to feed you. I wish you were here, though."

"I'll come over to the clinic tomorrow and we can shout at each other from a distance."

"Good."

They said their goodbyes and Lisa hung up gently. She stared at the phone, wishing she'd prolonged their conversation. She wanted her husband with a fierce longing that took her breath away. What was she doing in this crummy hotel without him? Suddenly the rendezvous with Greg seemed unimportant, superfluous.

And Greg himself had lost his luster. Well, wasn't it about time? They'd had their last adventure together, and now it was time for Greg to return to his nomadic existence in Israel and for Lisa to go home.

Home, with all that implied. Kids and cats, job and husband. It was, in every way, where she belonged.

She picked up the phone again and dialed James' number.

"James Barber here."

"I forgot to say 'I love you.'"

She heard him sigh. "I love you, too. Now go get some food, and try to stay out of trouble."

Lisa shoved the hair dryer back into its holder and rummaged in her suitcase for a clean shirt. Greg would have to take her as she was, bruised and rumpled, with wet hair and no make-up. She was too tired and hungry to primp.

Greg watched her shovel the late night dinner of lamb and couscous into her mouth. He grinned, admiring the efficiency of her eating, which was all about fueling the body without caring a jot about how she looked.

Lisa caught his look and smiled back. "What, haven't you ever seen someone eat before?"

"I was admiring both your appetite and your no-nonsense method of putting food away."

"I was hungry!" She pushed her plate away and sighed.

Greg noticed a certain lack of tension in the air. The sense of possibility between them—an almost tangible thing the last few weeks—had disappeared. The energy of the quest was dissipated. So be it, he thought. They were approaching the end of this particular chapter in both their lives. He tried to ignore the feeling of dismay this realization caused him.

"So, where are you on the Farouk family round-up?" said Lisa, who looked a bit frayed around the edges.

Greg lit a cigarette and inhaled gratefully. "Young Abdul will be hustled back to the clinic because he's still contagious. That is, when the health officials find him—he took off somewhere after I talked to him. Fortunately, his sister and mother were vaccinated soon enough, so they didn't get smallpox. Mustafa Farouk and Abdul's cousin Khalid are being questioned," he said. "But they have been unable to provide any information on the whereabouts of Zawi or Zawi's boss—someone in Israel or further north whose name we don't know." Not for lack of trying, either. Greg wasn't about to detail the combined methods of the CIA and Mossad for interrogating terror suspects.

"Is there really a Hamas connection?"

Greg nodded. "Oh, yes. And Hezbollah is involved, too." There was no point in holding back the essential outlines of the network since the media would be releasing the story any day now. He saw that Lisa was gazing off into space. "What are you thinking now?"

"About the little sphinx logo. How it's a copy of the Great Sphinx at Giza. It took me quite a while to remember that it was on an entrance wall of the Farouk house as well as on several packages. I suppose you've tracked down all the contacts the Farouks had in the Cairo area?"

Greg sighed. Lisa really would make a great undercover agent with her memory for little details. Too bad she was wedded to both the academic life and an American husband. "Funny you should mention that. I've had Rafi up there chasing down leads since we discovered the carpet seller who was disseminating smallpox dust on silk fibers."

His cell phone rang. Greg whipped it out of its pouch and listened.

"Yeah. I thought so…excellent work. We'll get right on it."

He flipped the phone shut and grinned broadly at Lisa. "You have an uncanny knack for timing," he said. "That was Rafi. He's talked to Abdul's sister and her fiancé, who were persuaded to describe several hidey-holes and meeting places the Farouks used in the Luxor area. They also knew of a couple of cafés used in Cairo. We'll get Zawi now."

"So Rafi followed the Sphinx, too?"

"Yup."

"Great minds think alike. Now, is it okay if I get some sleep?"

Chapter Twenty-nine

Thursday afternoon, September 16, Luxor

The next afternoon, greatly refreshed by a dreamless nine hours of sleep and a scalding, hot shower, Lisa descended the stairs of her hotel in search of sustenance.

She found the hotel dining room almost deserted since lunch was over, but at a table near the back of the room was a familiar profile.

Abdul, with a young woman who could only be his sister, Zara, and another young man she'd never seen before. All three were sipping Arabic coffee, facing Greg Manzur.

Lisa pulled up a chair on Greg's side of the table.

"Lisa, this is Abdul's sister, Zara Farouk, and her fiancé, Farzad Hammadi. Did you sleep well?"

"Like I'd been dropped into a black hole," said Lisa. She looked at Farzad and noticed his sling. "What happened to your arm?"

Zara spoke for him, her black eyes sparkling with anger and something else. "My father broke it. When he caught us trying to elope. And wounded him as well—with a knife. But now..."

"Now my father will not harm anyone for long time," finished Abdul with a triumphant grin.

Greg explained. "Based on the information these three young people were able to give us, we found Zawi Farouk hiding in a friend's house just south of Luxor. Zawi and his brother Mustafa are both in custody now, and we hope to make them talk." He smiled grimly and amended his own statement. "We will make them talk—we need much more information about the links to other parts of the Middle East. Where the virus stocks came from, who's funding the operation, what the other targets were."

Lisa stared at him. "Sounds like you've got your work cut out for you."

"Yeah. I'll be busy for months, maybe years."

She realized he was not exaggerating. Tracing the path of disease and destruction the terrorists had unleashed would take international cooperation, hard work, and luck.

Zara and Farzad were holding hands and whispering together.

Greg's face relaxed as he looked at the three expectant young people. "Now for more pleasant topics," he said. "I am authorized to pay you for the extremely useful information you have provided. You, Abdul, must take a test, but then you can enter the high school-to-medical-school special program at Cairo University in two months time. All your registration fees and living expenses will be provided."

Abdul's dark eyes shone. "I will become doctor. It is my dream."

"But first you need to go back to the clinic for a few days," said Greg.

"I am well enough! I don't want to go back there."

Greg skewered him with a look. "You must. You are still contagious to people who have not been vaccinated. You must cooperate with the health officials before I will let you go to Cairo."

Abdul's shoulders slumped. "Okay, I go back."

Greg turned to Zara and Farzad. "And you two will receive a sum in your new bank account in Aswan to set up housekeeping and open your own shop." He named the amount.

Zara's eyes grew huge and Farzad erupted with joy.

"Hamdul'allah! And thanks to your agency!"

"And to you!" added Zara, with a brilliant smile that made her look about ten.

"Don't mention it," said Greg. He winked at Lisa, and she smiled back.

Lisa guessed that he wasn't authorized to tell them the source of the funds. "When are you getting married?" she asked Zara.

"Sunday," was the reply. "And you are invited. My mother insists."

Greg asked Lisa to walk with him over to the Nile. She agreed, sensing he had something to tell her.

They reached the long walkway that ran alongside the boat quay where all the hotel boats were moored. Most were forlorn, empty of tourists and crews. Lisa reflected that the smallpox outbreak was going to put quite a dent in local incomes. Egyptians in Luxor, as in Cairo, depended heavily on the tourist trade to get through the winter season.

They leaned against the low wall, watching the turgid water flow by, adorned with clumps of hyacinth and occasional sesame rushes.

Lisa turned to find Greg watching her with mischief in his eyes. "Say, how about a little camping trip in the desert before you fly home?"

Instantly, she was transported to Ein Gedi, Israel, where she and Greg had camped when she was twenty. Warmth crept up her neck as she remembered the first time they'd made love—almost sixteen years ago.

"Cut it out, Greg. You're manipulating me."

"But it's so much fun to make you blush."

"Good grief! That was a million years ago!"

"And you remember it as well as I do," he said, the smile leaving his face.

Lisa realized the teasing had ceased. "What is it, Greg?"

"I have to go back to Cairo in an hour or so," said Greg abruptly. "Then back to Israel. I've been called back to try and trace the Hamas end of this mess."

"Then this is goodbye, isn't it?"

"Yes." His mouth twisted slightly. "I bet you aren't coming back to the Middle East anytime soon."

"No, I doubt it. James said he'd like to go on a safari sometime, but that means East Africa—or maybe South Africa. But that will probably wait until our kids are older."

They faced each other. Lisa's chest felt tight. Suddenly, it didn't matter that their relationship had never worked, that she'd known each time that he was the wrong guy for her, that she was happily married to someone else. What mattered was that she'd probably never see Greg Manzur again.

"I don't get over to the States much, either." He let out a whoosh of breath. "Okay if I email you occasionally?"

"Sure, but I'm a rotten correspondent when work at the museum heats up." She knew she'd write back, at least at first. But she also knew it was time for Greg to take his proper place—in the past.

Greg held out his arms and Lisa moved into them. They hugged, long and hard. Then Greg pulled back so he could look at her face. "You're a hell of a woman, Lisa. And—for the record—you'd make a great agent."

Lisa saw that his brown eyes were unusually bright and felt her own tears well up. She kissed him on the cheek and patted his shoulder. "Thanks, but no thanks. I'm too old for car chases, I hate rock climbing, and I'm sick of toilet paper. I'll stick to museum work."

"I understand," he said, and a rueful smile tightened the corners of his mouth. "Stay safe, okay, Lisa?"

A car pulled up next to them, Rafi at the wheel. "Boss? We gotta go."

"Goodbye." He hopped in the front seat.

" 'Bye. Take care of yourself, Greg."

Lisa watched the car roll away. She turned back to the soothing flow of the Nile and let the great river take her tears away.

Chapter Thirty

Friday morning, September 17, Luxor

"What was your reaction from the vaccine?" asked Mark Leonard.

"My arm got a little sore, then turned red. Then I got a little pustule which scabbed over," said Lisa.

"Excellent," said the young doctor, who had introduced himself as James' closest colleague and minder at the field clinic. "That means that even though you spent time with Abdul Farouk, who is still contagious, you are almost certainly protected from the disease. Just to make sure, we'll keep a close eye on you over the next two weeks. If you develop any symptoms, especially a fever or rash somewhere else on your body, you call me at this number, okay?" He handed Lisa a slip of paper.

She nodded. "How is James? I haven't talked with him today."

Mark smiled. "Much better. In fact, we've just changed the diagnosis. When he never developed more than two pustules—one on his forehead and one on his chest—we were rather surprised. When we questioned him, we found out about his eczema…"

"Why is that important?" asked Lisa, suddenly motionless in her chair.

"People who suffer from eczema often react very badly to the smallpox vaccine. The symptoms can mimic the disease, at least for the first few days."

Lisa's hands moved up to her face. "So he doesn't really have smallpox?"

"That's right."

She jumped to her feet. "When can I see him?"

Mark smiled broadly. "Now. He's on his way to your hotel."

A few days later, James and Lisa attended a CDC conference together.

James reveled in the feeling of his wife's hand clasped in his own.

He stole a look at her face as she sat next to him. Her skin was clear and her eyes faintly shadowed—he'd kept her awake for quite a while last night—but her expression was serene.

"Hey," he said softly, as the room filled up around them with doctors, nurses, and health officials.

Lisa smiled at him. "Hey, yourself. Feeling better?"

"What do you mean?"

"You know darn well what I mean," said his wife. "Even your snoring sounded satisfied."

Dr. Thomas Mallory tested the microphone. "Please take your seats. We'd like to get started."

The chattering in English, Arabic, French, and German subsided as people settled down.

Dr. Mallory cleared his throat. "Before I give you my report, I have a sad announcement to make. Our colleague from New York, Dr. Stanley Rosenberg, died early this morning."

James felt his vision go dark, and he squeezed Lisa's hand so hard that she whimpered. Stanley'd had a heart condition—that was probably a contributing factor to his death—but that knowledge was no comfort. What a great guy. What a bloody waste.

"So far we've had five deaths in Egypt and Israel, and twenty active cases that we know about. Fewer cases than we could have had, all things considered. That young fool, Abdul Farouk, has been slapped back into quarantine after going AWOL from the field hospital. We've got the Germans isolated, and only one Japanese tourist got sick. He was stopped at Cairo airport and put into quarantine."

"How did the terrorists disseminate the dust this time?" asked an Egyptian medic.

"Pillows."

"What?" said a doctor several rows in front of them.

"Well, actually they were cloth covers decorated with Egyptian designs," replied Mallory. "People buy them to hang on the wall or fill with pillows when they get home; the cotton fibers on the inside of each cover hold smallpox dust very nicely."

James remembered Lisa admiring such pillowcases. Blue and maroon ones, with stylized lotus plants and lozenges. What if she'd bought the pillowcases, taken them home to Emma and Sam? His hands, recently released from Lisa's clasp, clenched so hard that his fingernails dug into the skin of his palms. Think how many other families had already been affected. Twenty cases still active, meaning about thirty percent of them would die, leaving the rest disfigured.

Or blind. James'd had plenty of time to think about the consequences of smallpox recently. He could handle a few scars, but his mental and emotional well-being—not to mention his job—was completely dependent on his being able to see images: X-rays, CT scans, and photographs. Going blind was his worst nightmare.

A reporter called out, "What methods are being used to contain the outbreak?"

"Same as in the 1970s," replied Mallory. "Surveillance-and-ring vaccination. We swoop in and vaccinate everyone who had any contact with the victims until the virus has no place to go."

"Could you review that for us?" asked the same man.

"Okay. Well, the original Smallpox Eradication Program succeeded because variola only has one host: humans. So if you vaccinate all the people around an outbreak, the variola finally dies out. The story of the vaccine is quite interesting..."

James felt a little better as Mallory's focus shifted from the human toll of the disease to its more intellectual aspects. Mallory reviewed the development of the vaccine from cowpox, a form of pox related to vaccinia (a milder form of smallpox) that passes from cows to people. Physician Edward Jenner noticed in 1798 that dairymaids who had contracted and recovered from cowpox were then immune to further cases of cowpox—as well as smallpox. Jenner's first vaccine (from Latin "vaca," or cow) was fluid from cowpox lesions that he scratched into the skin of healthy people, who then developed immunity to smallpox.

"In 1977, a Somalian cook broke out in the last natural case of variola. It took fifty-seven thousand vaccinations of people around him to stop the outbreak," continued Mallory.

"How many vaccinations have we done this time?" asked a young CDC woman.

"Fifteen thousand and counting," said Mallory. "But the prognosis is better than it was for American Indians in the 1600s. They had no immunity and no response plan. We have the advantage of knowledge and a vaccine that can limit the spread of the disease if it's administered quickly. It's still going to cost a helluva lot of money and man hours, though."

"Did anyone get on an airplane before he was vaccinated?" asked James.

Dr. Thomas Mallory looked at the piece of paper someone handed him and his lips tightened into a thin little line.

"Yes," he said quietly. "I just received notice that a Swedish tourist flew home to Oslo and infected her entire family. The eradication team is on its way to Europe, and the vaccinations of everyone who had any contact with the woman will begin in the next twenty-four hours."

Chapter Thirty-one

Friday afternoon, October 10, Cairo

"Dried or frozen smallpox preserves very well. The weaponized virus left over from the Russian experiments can be kept safe in a remote location for many years."
(From the diary of Stanley Rosenberg, M.D.)

"Kind of nice to be just a tourist again," said James as he squeezed between café chairs on the sidewalk.

"You mean, as opposed to being a patient in a field hospital or an assistant Mossad agent," replied Lisa.

"Yeah. At least you have the remains of a tan. I look as pale as if I'd never been in Africa. And it's October now—it will be cooler in Boston."

They were enjoying their last half-day of freedom before being imprisoned in the narrow rows of an overseas plane.

"There it is," said Lisa, pointing at a throng of people jostling under strings of lights and racks of hanging clothes. "We can't leave Cairo without going into the Khan al' Khalili."

James grimaced. "I'm not sure I have the energy for this," he said.

"Half an hour, tops. I promise," said Lisa.

The most famous bazaar in Egypt sucked them into a vast labyrinth of crowded stalls. Tourists from every country mingled with smooth-talking vendors with flashing eyes and gold-toothed smiles.

"No charge for looking!"

"This way! You are going much more better this way!"

She and James ducked under a colorful Ramadan lantern and entered a narrow, cobbled alley off to the right.

Lisa admired a stall of brass dishes and coffee tables with hammered designs until James tugged on her shoulder.

"Look—the same sort of pillow covers that were used in one of the attacks."

She followed his pointing finger to a stack of maroon and navy appliquéd covers. Instantly, the well-dressed vendor picked two up to show her. "You like?" he asked.

"No thanks," said Lisa, moving away with a slight shudder.

The colors and shapes of so much merchandise were overwhelming. Gold and silver chains, with and without the ankh symbol of eternal life, fake Pharaoh's heads, plaster cats, rugs of silk and wool. Colorful drawings on "authentic" papyrus, complete with a certificate that was supposed to give you confidence that you were not buying cheap paper made from banana leaves. Chessmen, made out of ivory or alabaster, and inlaid boxes of precious wood. Lisa marveled at the high quality art works mixed up with kitsch and cheap souvenirs.

"We haven't had much time for shopping on this trip," said Lisa. "Anything you really want?"

"How about a leather shop?" he said, surprisingly. "I could use a new belt."

In seconds, a merchant appeared at their side, urging them up a narrow flight of stairs to his shop. James and Lisa allowed themselves to be led.

While James fingered various belts, Lisa decided to play it cool and let James do the bargaining.

"How much is this?" he asked, holding out a narrow navy belt with an attractive stamped design and a simple silver buckle.

The vendor named a figure that Lisa quickly converted to seventy American dollars. She shook her head at James.

"Too much. I can get a belt for less at home," said James.

"Ah, these are the best in all Cairo!" said the man smoothly. "Tell me what you want to pay. Between you and me, my friend, we make a good price."

Lisa listened to them dicker, distracted by the sight of a lean, white cat padding over some rugs in the stall next door.

James moved towards the door in the time-honored manner of an experienced bargainer.

Changing his tactics, the vendor spoke to Lisa. "Madam! You choose belt for husband."

Lisa smiled and shook her head. "No, he must decide."

"Madam, I think your husband must be very rich man—he is so reluctant to part with his money!"

When Lisa and James had escaped down the stairway to the main thoroughfare below, they shared a good laugh.

"You did that awfully well, James," said Lisa, still chuckling.

"I did want that belt," he said. "But I was convinced he was overcharging even at the final price."

"No doubt. They think all Americans are just walking wallets."

Lisa took his hand and they strolled along, enjoying the jumble of people and languages, steadily resisting the efforts of vendors to draw them in unless they were truly interested. In the end, Lisa purchased a small rug with an exquisite tree-of-life design and James found a little cat statue for Sam.

Then they walked back to their hotel.

"You should go, Lisa. I'm beat." James dropped his daypack on the floor and stretched out on the king-sized bed. "Besides, it's your friend, and you'll want to talk shop."

"Are you sure? I hate to leave you alone on our last evening in Cairo," said Lisa. She was dying to see Jeff and Alice again—this was her last chance, maybe forever, to see her archaeologist friend and his wife.

"Yeah, I'm sure. I'll have a nice snooze and a shower while you're off gallivanting. But leave me Jeff's phone number, okay?"

"You mean, you think I might disappear into the streets of Cairo and never be seen again?"

"Stranger things have happened. I hope you're going to ask the hotel to find you a reliable taxi driver?"

"Don't worry, I'm not climbing into any vehicle until I get assurances on both destination and price," Lisa said.

James accepted a quick kiss from her and then punched the pillow into a more comfortable shape and closed his eyes. Before Lisa had found the number and connected with Jeff for directions on how to get there, a gentle snore emerged from the region of James' bushy beard.

The hotel staff found her a forty-something Cairo native who made the Sheraton one of his regular stops. After watching the bellhop give the driver directions and seeing his nod of understanding, Lisa climbed in the front seat of the taxi.

The communication up to that point had been entirely in Arabic. Lisa discovered the driver's knowledge of English was extremely limited, but he was a good driver. His skill in easing out into traffic between parked cars and navigating without lanes—or headlights, except when he wanted to pass—was nothing short of amazing.

Lisa, exhausted by the complicated journey from Luxor to Cairo, suspended her usual caution and relaxed. A little voice said, "this is dangerous," but her tired brain said, "I don't care—I'm going to see my friends. We're leaving the country in a few hours, and I'm damned if I'll miss this chance."

Her reward was exhilaration. Lisa loved the speed of flying through the streets of a largely unknown city at night. It made her feel powerful to rely on only herself and her three or four words of Arabic. And when the driver stopped for cigarettes, Lisa took over, unwrapping the package and lighting one for him so he could drive with two hands on the wheel.

Together, Lisa and the taxi driver got lost near the Grand Mall, asked for directions using a comic mixture of pantomime and her scrawled and totally inadequate map—a couple of squiggles on a hotel pad. Rescue came in the form of two passing youths who helped her dial Jeff's phone number on a cell phone that had numerals in Arabic script.

It felt like a minor miracle when they reached Lisa's destination, an obscure apartment building in the suburb of Mahdi. She enjoyed a ridiculous feeling of triumph, because somehow that taxi ride, all on her own, made her feel like a carefree student again.

Telling Jeff and Alice the saga of her disease-and-danger ridden trip to Egypt was therapeutic. Lisa's friends listened with appropriate sympathy and horror, adding insights from their ten years living as Americans in Cairo. After feasting on roast chicken and salad, Lisa and Jeff had indulged in some archaeological gossip while Alice dished up some homemade yogurt with honey.

Jeff leaned forward, eyes gleaming. "Do you remember Sally Dickerson? She was a curator at the MET, and she moved over here and took the job at the Cairo Museum."

"Sure I do! How's she doing?"

"She's turning the curation policies upside down, for the better, of course…"

Later, the taxi driver, who had waited for her, whisked Lisa back to her hotel. He accepted her thanks and her generous tip with a wide grin and vanished into the streets of Old Cairo. She looked at her watch and smiled. Her adventure had taken only three hours and she was back in plenty of time to catch their three a.m. flight home.

Lisa's nose was pressed against the thick glass of the airplane window as the lights of Cairo and the continent of Africa fell away below them. Beside her, lay James, dozing as the plane climbed higher.

She'd had a little too much wine at Jeff's, which made her mouth dry and her thoughts disconnected. Her mind skittered back and forth between the pleasant memories of her evening out and the horrific results of the terrorist plot to kill Westerners with a deadly virus.

Fifty-two very sick tourists—and that was just in Egypt. Nineteen deaths, four in her group—Stanley Rosenberg, Nancy Keller, Betsy

Grover, and that cute little boy, Nicky Reese. Wasted lives, especially the seven-year-old's. Lisa had really liked ditzy Nancy, with her inane questions and cheerful demeanor. She'd gotten the smallpox from handling contaminated carpets in Cairo. And Betsy, the intrepid ER nurse. She felt guilty as well as sad—guilty that she'd been jealous of Betsy, and guilty that she and James had made it when so many others had not. Especially when she thought of Nicky's mother, Mrs. Reese. How horrible it would be to lose Emma to such a disease.

She slept a little. Woke and dozed again, her eyelids fluttering. Her motionless body felt glued to the seat but the video camera in her brain continued to generate random images.

The Great Sphinx, echoed in the terrorist logo on brown paper packages. The smallpox scars on Abdul's face. The filthy, uncomfortable room full of toilet paper that had been her prison.

Lisa opened her eyes and stared out the window. Bright stars winked in a black ice sky.

She sent good vibes to Greg Manzur back in Israel, chasing down Hamas connections. And to Thomas Mallory, still mopping up after a huge operation in Egypt and communicating with health officials across the globe. They would have to eradicate smallpox all over again, at hideous expense.

She thought of her original expectations for this trip—an exotic vacation punctuated with archaeological wonders, little adventures, and intimate moments to remember fondly in the future. Instead, she'd spent most of her "honeymoon" being worried, sad, or terrified. She'd almost lost James—and most of her memories were too horrible to save in a scrapbook.

Lost—and found. Lisa turned to look at her sleeping husband. What she'd found, despite everything, was better than any honeymoon. Her marriage was more than a pleasant backdrop to the rest of her life; it belonged at center stage. Everything that mattered to her—children, job, making her small mark on the world—involved James, and she was unspeakably grateful that he was alive, unharmed, and right next to her.

The stewardess pushed the drinks cart further down the aisle beyond their seats, and James woke up and stretched.

"Ready for boring old Boston?" she asked, feeling a surge of longing for her own city and her place in it.

"It won't seem boring to me!"

"I can't wait to see Emma and Sam—and Oreo."

James laughed. "You would include that darned cat! But I feel the same way. Too bad we have to wait twenty hours or so to see them."

Lisa leaned against James' shoulder and wished for a magic carpet or Dorothy's ruby slippers.

There's no place like home.

Acknowledgements

Once again, I am indebted to friends and family for their support in writing this novel. The inspiration for the story was our trip to Egypt in September 2005, but the addition of bioterrorism to the original archaeological plot was my husband Charlie's idea. He also contributed medical information and travel details.

Huge thanks to my writing partner, Molly MacRae, who read two drafts for character development and inconsistencies, and to Pat Ricks of Pemberley Press for additional suggestions. Their input improved the manuscript enormously.

I also thank Shirley Jensen, Ottilia Schershel, and Jo Ann Temple Dennett, members of my online Guppies critique group, for numerous helpful comments and corrections. And I thank Anna Merritt for her excellent proofreading.

Below are some of the sources I used for background reading and chapter heading quotes:

Ethne Barnes, Diseases and Evolution (University of New Mexico Press, 2005)

Richard Preston, The Demon in the Freezer (Ballantine Books, 2002)

Abigail A. Salyers and Dixie D. Whitt, Microbiology: Diversity, Disease and the Environment (2001). Online supplement, "Rapid Response Chapter: Bioterrorism" (Fitzgerald Science Press, Inc., 2002)

Willis Roth Regier, Book of the Sphinx (University of Nebraska Press, 2004)

James B. Prichard, The Ancient Near East: An Anthology of Texts and Pictures (Princeton University Press, 1958)

Website for the Centers for Disease Control Smallpox Response Plan: www.cdc.gov

Sarah Underhill Wisseman writes archaeological mysteries starring Lisa Donahue, an archaeologist and museum curator. Sarah hadn't a clue that she wanted to be an archaeologist until she traveled to Israel right after her freshman year in college. There she ate felafel, fell in love with Jerusalem, camped illegally on Masada, and spent a month at the excavation of biblical Beersheba. Once hooked by archaeology, she returned for her junior year at Tel Aviv University, an experience that changed her life.

Back in the U.S., she completed her B.A. in Anthropology at Harvard University and volunteered in the Harvard Semitic Museum and the Peabody Museum in Cambridge, Massachusetts. Then she left her home state of Massachusetts and moved to Philadelphia for graduate school in Classical and Near Eastern Archaeology at Bryn Mawr College. There she met a medical student, Charlie Wisseman, who became her husband in 1975.

Sarah had just finished her M.A. and was working on her Ph.D when Charlie landed his first job as a pathologist working for the Public Health Service in Cincinnati. She seized the opportunity to learn more about art conservation by volunteering in the laboratory at the Cincinnati Art Museum.

When her husband migrated to a job in Urbana, Illinois, Sarah became a curator, grant writer, and fundraiser at the World Heritage Museum, located in a creepy, fourth-floor attic at the University of Illinois. Somehow she survived the experience of working there while raising two children, Nick and Emily. Then Sarah was lured away from the museum world into archaeological science. She currently directs the Program on Ancient Technologies and Archaeological Materials (ATAM), also at the University of Illinois.

The House of the Sphinx is Sarah's fourth mystery. The first novel, Bound for Eternity (2005), was a finalist in the 2004 St. Martin's Press Malice Domestic contest for Best First Traditional Mystery. Although set in Boston, this book is based upon actual research on Egyptian mummies at the University of Illinois.

Sarah's second novel, actually a prequel to Bound, was The Dead Sea Codex (2006). In this suspense story, curator Lisa Donahue and another archaeologist race to find a lost manuscript in Israel before Christian fanatics destroy it. Lisa's career continues in The Fall of Augustus (2009), when a falling statue kills her boss and Lisa becomes Director of her Boston museum.

Sarah currently lives in Champaign, Illinois, with her husband and two cats. A live interview with Sarah can be found at "Voices in Mystery" on YouTube.

For more information about her short stories and articles, visit www.sarahwisseman.com

Made in United States
Cleveland, OH
05 November 2026

47664331R00114